NARRATIVES IN MEGAPROJECTS

This book is a novel contribution to a field dominated by conventional approaches to project management; it is about narratives in megaprojects. Among the questions examined in this original new book are:

- What are narratives?
- Why are they important in megaprojects?
- How are they formed and used in megaprojects?
- How do promotors of and protestors against megaprojects craft narratives to their advantage?
- What strategies can project managers employ to effectively use narratives in megaprojects?

Built from longitudinal research studies in combination with internationally recognised teaching materials, this book will provide readers with a theoretical understanding of narratives and projects, as well as practical international case studies, including HS2, the Dakota Access Pipeline, the Eden Project and Thames Tideway, to support their understanding. The authors explain the different types of narrative, and how and why they are important in general and in relation to a megaproject and its lifecycle, but also explore how to craft narratives in different situations, and how they are changed and maintained over a project's lifecycle.

Narratives in Megaprojects doubles as a text supporting more advanced courses on project management or aspects thereof, and as a reflection of the state of the art in this particular perspective on megaprojects. It is essential reading for all students and professionals in project management, construction and infrastructure as well as executive leaders involved in megaprojects and infrastructure delivery.

Dr Natalya Sergeeva is an Associate Professor at the Bartlett School of Sustainable Construction, University College London (UCL). Natalya lectures on project management and innovation management at the postgraduate and executive levels. She has some practical experience managing construction and infrastructure projects and consultancy. Natalya's research explores the nature and role of narratives in leading projects and firms, individual and organisational identities, and the ways leaders articulate and translate narratives and identities. She has published a number of articles in leading journals, such as *Industrial Marketing Management*, *International Journal of Project Management*, *Project Management Journal*, *International Journal of Innovation Management*, and *Creativity and Innovation Management*.

Dr Johan Ninan is an Assistant Professor at the Faculty of Civil Engineering and Geosciences, Delft University of Technology (TU Delft), the Netherlands. Previously, he was a Post-Doctoral Fellow in the Bartlett School of Sustainable Construction at University College London (UCL). His research focuses on megaprojects, stakeholder engagement, collaboration, innovation, and project organising with a particular emphasis on the role of digital media. He has published in leading project management journals such as *International Journal of Project Management*, *Project Management Journal*, and *Construction Management and Economics*. He was awarded the PMI Young Researcher Award, the IPMA Global Young Researcher Award and the APM Paper of the Year Award.

NARRATIVES IN MEGAPROJECTS

Natalya Sergeeva and Johan Ninan

Routledge
Taylor & Francis Group

LONDON AND NEW YORK

First published 2023
by Routledge
4 Park Square, Milton Park, Abingdon, Oxon OX14 4RN

and by Routledge
605 Third Avenue, New York, NY 10158

Routledge is an imprint of the Taylor & Francis Group, an informa business

© 2023 Natalya Sergeeva and Johan Ninan

British Library Cataloguing-in-Publication Data
A catalogue record for this book is available from the British Library

Library of Congress Cataloging-in-Publication Data
Names: Sergeeva, Natalya, author. | Ninan, Johan, author. Title: Narratives in megaprojects / Natalya Sergeeva, Johan Ninan.
Description: Abingdon, Oxon; New York, NY : Routledge, 2023. |
Includes bibliographical references and index. |
Summary: "This book is a novel contribution to a field dominated by conventional approaches to project management, it is about narratives in megaprojects. It doubles as a text supporting more advanced courses on project management or aspects thereof, and as a reflection of the state of the art in this particular perspective on megaprojects"-- Provided by publisher.
Identifiers: LCCN 2022047298 | ISBN 9781032163963 (hbk) |
ISBN 9781032160139 (pbk) | ISBN 9781003248378 (ebk)
Subjects: LCSH: Civil engineering--Information services. | Construction industry--Information resources management. | Construction projects--Management. | Project management.
Classification: LCC TA158.5 .S38 2023 | DDC 624.068--dc23/eng/20221206
LC record available at https://lccn.loc.gov/2022047298

ISBN: 978-1-032-16396-3 (hbk)
ISBN: 978-1-032-16013-9 (pbk)
ISBN: 978-1-003-24837-8 (ebk)

DOI: 10.1201/9781003248378

Typeset in Bembo
by SPi Technologies India Pvt Ltd (Straive)

CONTENTS

CASES

PREFACE

Our motivation for this book is to provide an original theoretical and practical contribution to knowledge into megaproject management discipline. We will cover what narratives are and why they are important in terms of their implications for theory and practice. The term 'narratives' is used in different contexts and in different disciplines: politics, economics, project management, social science. We highlight how narratives are coherent as they bring together disparate experiences, are performative as they shape and change the future, are strategic as they can create, maintain or disrupt institutions, and are promotional as they are consistently communicated to brand an image of the organisation. The book consolidates our emergent research into the role and nature of narratives in projects.

The book is based on findings from the ESRC-funded research grant. Drawing on our longitudinal research into project narratives, we have published a number of articles, book chapters and case studies. These materials are used in our teaching of "The management of construction firms and projects" module for the MSc Construction Economics and Management programme at University College London (UCL). We hope that the materials in this book will be a valuable source for teachers and students worldwide. We intend this book to be a guide for students and researchers as it will contain theoretical understanding of narratives and projects, as well as practical case studies to support the theoretical understanding. It will also be a useful source for practitioners who are interested in learning more about narratives in projects, and tools and instruments available on how craft and maintain project narratives in their practices. The readers will learn about what narratives are, how they are used, how these can be crafted and maintained. For example, narratives about the desired future play an important role for branding a project. Whilst there are often counter-narratives in megaprojects, crafting a coherent and consistent narrative about a need for and benefits of a megaproject is pivotal. The readers will also learn about the implications for project narratives in actions. For example, the role of narratives in crafting project identity or project DNA. This book discusses narratives of future and ideas for future project narrative work.

The book will offer an insightful understanding about what narratives are, and how and why they are important in general and in relation to a project and its lifecycle. The book will teach how to craft narratives in different situations and what makes a project narrative. It will describe different types of narratives, and how they are changed and maintained over a project lifecycle. It will provide novel ideas into project narrative work in practice and thinking about future narratives. The global case studies provided in the book demonstrate how project narratives are crafted, used, managed and maintained in practice. We believe and hope that you will find our book inspirational.

1

WHAT ARE NARRATIVES AND WHY THEY ARE IMPORTANT?

1.1 Definition of a narrative

What is narrative? In this book we define narrative as a discursive construction that embodies unity of purpose, a degree of coherence together with connotations of performative intent. Narratives are defined as temporal discourses that provide "essential means for maintaining or reproducing stability and/or promoting or resisting change in and around organizations" (Vaara et al., 2016, p. 496). They are a unique form of discourse. In this book we distinguish narratives from stories, although some organisation scholars have used these terms interchangeably. Narratives are characterised by coherence, performative intent and repetition, whereas stories are more personalised, entertaining and emotional in nature. In contrast to narratives, stories are understood as relatively fleeting idiosyncratic accounts of personal experiences. Both narratives and stories are important in structuring the lives of people and the actions of organisations.

Narratives are widely accepted as a vital means of organising (Currie & Brown, 2003; Rhodes & Brown, 2005). They help to create an order out of disparate experiences, events and situations. They are instrumental in organising the world we see around us by embodying coherence and unity of purpose (Pentland & Feldman, 2007) and are also termed as "the style and substance of life" (Trible, 1984). There is undoubtedly a strong and increasing interest amongst scholars in narratives and their importance in relation to various disciplines and contextual meanings. The multidisciplinary research on narratives includes subjects such as economics, politics, sociology, organisation and business management studies, strategic management, innovation management and project management.

In the past, narratives have mainly been studied as a method of research and data analysis, being often referred to narrative inquiry methodology or narratology. The research on narratives has been expanded in recent years. The narrative interview

DOI: 10.1201/9781003248378-1

method has also been introduced into organisation studies, innovation and project management research. Some recent research adopts a narrative perspective on understanding new emerging concepts such as innovation, sustainability, climate change, digital transformation, safety, health and wellbeing. This book is a theory of narratives and practical application into project management discipline and has broader implications for multidisciplinary research and practice. The international case studies in this book demonstrate how narratives are used, along with their importance and implications. It provides guidance on how narratives are crafted, maintained and changed in the context of global megaprojects.

The 'narrative turn' in organisation studies constitutes a shift in focus away from the material and descriptive practices towards understanding how the meaning is socially constructed. Narratives are important vehicles through which meanings are negotiated, shared and contested. Narratives are about meaning-making and they have to be crafted according to their context. They are attempts to ascribe events and happenings with a meaningful order. Boje (2001) argues that narratives are characterised by the possession of a plot together with a degree of coherence. Organisations need coherent narratives because narratives represent attempts to impose order.

1.2 Key distinguishing characteristics of narratives

This book extends the 'narrative turn' in organisation studies and project management research. It distinguishes narratives from stories and other types of discourses through their key characteristics outlined below:

1.2.1 Coherent

Narratives tend to be crafted in a logical order in organisations, so that they make sense for people. Narratives are hence integrated as a whole and become coherent and consistent over time. They bring coherence to disparate experiences, events and situations. Organisations are full of contradictive views from people both inside and outside these organisations, and people often make sense of these contradictive views through narratives. Narratives do help to generate coherence and consistency, and hence crafting such coherent and consistent narratives becomes important practices in organisations. Narratives seek to bring plausibility and coherence to disparate and shared experiences (Cunliffe & Coupland, 2011). Managers are expected to craft and communicate coherent narratives about their organisations and organisational performance to their employees and wider audience (Sims, 2003).

1.2.2 Performative

Narratives are performative in nature because they potentially shape and change the future (Sergeeva & Winch, 2021). Performativity of narratives entails change in organisation by creating action and, in turn, results in new narratives. For example, outdated narratives become updated or new narratives are crafted in response to new challenges and emerging situations. Narratives can be captured in formal texts, in particular, in strategy and speech texts, which in essence have performative intent and generate future actions.

Performative narratives are used for different purposes in organisations such as to achieve shared meaning and understanding, create and reinforce messages, and convince people. Narratives are seen as performative with an explicit intention of persuading the audience and delivering strategies (Hansen, 2012). Performative narratives also invariably play an important role in legitimising advocated actions (Frandsen et al., 2017).

Narratives are characterised by "persuasive power of narrative repetition" (Dailey & Browning, 2014, p. 27). Narratives' repetition in organisations serves to stabilise particular meaning. For example, narratives can be used to achieve a shared understanding about different labels and terms used in organisation; they can also be used to achieve a consensus from contradicting and conflicting narratives. Although temporary, rarely fixed or completely monolithic, narratives are nevertheless often repeated in organisations (Dailey & Browning, 2014). Performative narratives hold the power to shape organisational and individual worldviews and mindsets. Yet even dominant narratives can be challenged and criticised.

1.2.3 Strategic

Narratives are often defined as strategies (Humphreys & Brown, 2002; Rhodes & Brown, 2005; Vaara et al., 2016). Actors who create, maintain or disrupt institutions often rely on narrative devices to do so (Lawrence & Suddaby, 2006). They are used purposefully in organisations and hence service as strategies. Visioning future-oriented narratives are strategies for organisational and individual directions and planned actions. Narratives become formalised in a form of organisational and industrial strategies to achieve long-term and overall goals and targets. Strategic narratives are crafted by managers to attain organisational and individual goals and achieve the desired state in the future. Narratives enable to craft a vision for long-term or overall goals, interests and the means of achieving them. Policy makers, managers and other practitioners use and communicate narratives strategically to convince investors and followers, to reinforce key messages, to form a common vision for a future.

1.2.4 Promotional

Narratives are consistently promoted in organisations by managers and teams to create a shared meaning that have important implications for organisational strategy. Narratives provide essential means in any kind of promotion in and around organisations, and sustaining them over time. Narratives tend to be deliberately used to create desired future, persuade audiences, legitimise actions and promote past achievements and other stories of success. Coherent promotional narratives about brand image are crafted and consistently communicated to external stakeholders in organisations (Ninan et al., 2019). Organisational narratives tend to be targeted at specific audience, either investors, clients, internal or external stakeholders.

Thus, narratives are coherent as they bring disparate experiences, events and situations together, are performative as they shape and change the future, are strategic as they can create, maintain or disrupt institutions, and are promotional as they are consistently communicated to brand an image of the organisation.

1.3 Forms or narratives

Narration and storytelling are used purposefully to convince audiences with messages and to entertain them when rehearsed narratives become repetitive. Chief Executive Officers (CEOs) and managers play an active role in crafting narratives, as being responsible for formulating and disseminating an organisational vision and strategies (Sims, 2003; Sonenshein, 2010). People may craft and communicate narratives for self-promotions and promotions of the work they have done or doing, as well as sharing their experiences with other audiences. For example, practitioners share stories about projects they work, photos and videos via social media (e.g. Facebook, Instagram, LinkedIn, Twitter) and share their perceptions and views in blogs and posts. Even users of a project share stories of their experience including the benefits realised from the project (Mathur et al., 2021). Promotional narratives may help people to win next projects, expand their network and create new opportunities.

Narratives are often spoken, but there are other forms of performed narratives such as written and symbolic/visual. These are often reproduced on policies and reports, corporate websites, or in other externally facing marketing material. Narratives can be corporate objectives in CEO statements, word of mouth circulated among stakeholders, academic publications discussed in seminars, or daily news media articles covering significant events (Ninan et al., 2022). Narratives in oral mode is shared in everyday conversations between employees, customers and suppliers (Hjorth & Steyaert, 2004). In contrast, scripted narratives are more formal and are used in formal presentations to external audiences, such as media representatives, bankers and venture capitalists (Martens et al., 2007). Narratives in written mode appear on promotional materials, such as company brochures, websites, product packages, annual reports, business plans or IPO prospectuses (Czarniawska, 1998). During the narration there should be an identifiable voice (Pentland, 1999; Pentland & Feldman, 2007). People do not simply tell narratives; rather, they enact them by putting them into practice. Voiced and enacted narratives may contain a sequence of events or discursive representations that embodies coherence or unity of purpose. Figure 1.1 shows the types of narratives and examples.

Textual or written narratives	Spoken or verbal narratives	Visual or symbolic narratives
- Published articles, reports, policies, strategies, newspapers - Written blogs in social media - Other textual materials and documentations	- Speeches, talks, presentations, webinars, interviews, conferences, conversations - Discussions and dialogues	- Videos, visual presentations, pictures, photos, graphics, music, illustrations - Other marketing promotional materials

FIGURE 1.1 Forms of narratives and examples

Narratives in their various forms carry important messages at the level of the organisation and wider. It is often a combination of these various narratives that provide coherence, performativity and consistency in organisations.

1.4 The importance of narratives and their implications

Narratives and narration play an enormously important role in organising by connecting the present with the future. They are the essential means for maintaining or reproducing stability and for promoting or resisting change in and around. Narratives organise people's lived experiences and create order out of random incidents, events and experiences. Such an order can help them understand the passage of events and even guide action. Different constructs, such as innovation, sustainability, digitisation, value and wellbeing at the individual and the collective levels, are social constructions often created by discourses and narratives. These narratives become popular and get repeated in organisations and could become dominant or grand narratives. Organisations can be better understood as continuously reconstructed entities with narratives driving practices and people's experiences. Organisational change process can be understood through narratives and a process of narration of negotiated meanings and enable to comprehend different employee voices. The case in this chapter, Fisht Sochi Olympic Stadium megaproject, shows how a narrative about environmental sustainability has been crafted and also the way in which it has been challenged. It also discusses a project narrative about cost overrun.

People make sense of their lives via narrative thought as the temporal and dramatic dimension of human existence is emphasised in them (Polkinghorne, 1991). The way people organise is dependent on the cues emanating from external perpetual senses and cognitive memories; and narratives are one of the main cognitive organising processes which shape temporal events around people. Hence, narratives are a "cognitive instrument" as they impact on the subject's thinking and emotional life. Strategically, policy makers and practitioners employ plotted, plausible and repeated narratives to shape the reaction of people to the changes occurring around them.

Narratives help make sense of organisational practices. For example, the innovation process in organisations can be understood through the ways in which managers speak, communicate and converse about innovation in the context of everyday practical activities. Establishing an identity, brand image and packaging organisations as innovative, sustainable and digitally transformed have become important narrative strategies in contemporary organisations. Narratives can make sense of the process of sustaining change in organisations by enhancing self-legitimisation. The future-oriented projecting nature of narratives means that they potentially shape and change the future (Sergeeva & Winch, 2021). They are performative in the sense that they are words (and other media) that impact decisions and actions we make. Narratives are persuasive in nature and are used by leaders to convince followers.

Narratives, and the process of narrating, have important implications for identity work. Organisational identity is about the meaning and understanding attached to the question: Who are we as an organisation? Organisational members and teams are expected to have a common shared understanding and an organisational identity where they work. An organisational identity narrative is conveyed internally in organisations. Identities of organisations are best regarded as produced by continuous processes of narration in which the narrator and the audience formulate, edit, applaud and refute various elements of the narratives (Brown, 2006). An organisational identity narrative evolves over time, as well as the variety of personal and shared narratives. Efforts of managers to control processes of organisational identity formation are often seen as hegemonic acts required for legitimation purposes (Humphreys & Brown, 2002). Establishing and maintaining an organisational identity has become an important part of organisational life.

Identity is a social and cultural phenomenon that encompasses macro-level categories, temporary and interactionally specific stances, and cultural positions (Bucholtz & Hall, 2005). Organisational identity is commonly understood as an organisation's members' collective understanding of the features presumed to be central and relatively permanent, which distinguish the organisation from other organisations (Gioia et al., 2000). Identities at all levels, be it at industry level, organisational level, project level, group level or individual level, can be considered as a social construction and they are subject to multiple interpretations and crafted by narratives (Stets & Burke, 2000). Narratives reflect and produce processes of identification in discourse. They help organisational members to dis-identify with the old and re-identify with the new identity. Theories, such as social identity theory, help people to understand how narratives employed by practitioners are used to identify projects, for example as innovative. Leaders consider it important to construct a coherent and consistent narrative, which in turn establishes individual and organisational identities which are instrumental for maintaining stability in the context of the ever-dynamic project environment (Sergeeva & Winch, 2021).

Individual identity or self-identity refers to subjective meanings and experience, to individual ongoing efforts to answer the questions: Who am I? and How should I act? Experience is understood as beliefs, meaning, norms and interpretations

(Kärreman & Alvesson, 2004). Individual identity is portrayed as fluid and fragmented, being frequently characterised by struggles and ambiguities. Individual identities necessarily draw on available social narratives about who one can be and how should they act (Thomas & Davies, 2005). Experience is constituted through individual narratives about self, others, and what has happened, and is happening to them. People act in certain ways on the basis of their memories, present experiences and expectations from available social, public and cultural narratives. Even when an individual constructs personal identity, he or she is embedded in socio-material contexts where individual thoughts and actions are shaped by other actors, work settings and broader social networks (Alvesson et al., 2008). An individual may construct a preferred (and often positive) narrative of identity to themselves and others. They may have several socially constructed labels attached to his or her identity, sometimes complementary, sometimes contradictory (Tomkins & Eatough, 2013). People may re-think and re-label their identities during their lives, as their beliefs, perceptions and social circumstances change. Individuals may re-construct their identities through contested and often-conflicting narratives about their everyday experience. Narratives about self, others and organisations are seen as primary means for crafting identity, while ongoing struggles and ambiguities are seen to be central in the process of identity construction. A process of identity construction may be constituted by several, more or less, complementary or conflicting identity labels. This emphasises processual aspects of identity constructions through narratives.

Narratives and narration have important implications for crafting and maintaining an organisational image or branding. Image is the perception of the organisational purpose, aims and values, resulting in the general impression in the mind of all stakeholders (Gregory, 2004). A positive organisational image creates trust and commitment in the members and help achieve a sustainable structure for the organisation (Kalkan et al., 2020). Narratives can be mobilised for creating a collective brand image at the levels of the firm and sector as a whole. Organisations communicate narratives through social media to create a brand image as beneficial for the community resulting in their support (Ninan et al., 2019). Organisations attract attention from wide audiences (through social media, external stakeholders, new media), which generates a need for a coherent narrative about the organisational brand image that is created and consistently communicated to external people. Promotional narratives play important role for crafting and maintaining an organisational image. They are temporal in nature and are required to be continuously updated to respond to the global trends in contemporary dynamic work environment.

1.5 Types of narratives

We can distinguish between different types of narratives. Narratives can be classified in terms of their levels, their temporality, their themes and as ante-narratives and counter-narratives. Each of these are discussed below.

One way of thinking about it is in terms of different levels. There are dominant narratives that can be identified at the global level. One such important dominant narrative is about sustainability and associated 17 Sustainable Development Goals (SDGs) developed by the United Nations. There is a widespread narrative about digital transformation of industries and digital technologies acting as an enabler of sustainability. Narratives of sustainability and digital transformation are connected with each other and could complement each other resulting in joint practices and activities. These narratives are connected with the broader narrative about challenges of climate change which affects everyone globally. The narrative about the importance of people's health and wellbeing is promoted globally in different ways, including government and industrial reports, firms' strategies and work practices. Narratives get established and crafted, maintained and changed, and get promoted by multiple global organisations such as the United Nations, government, political parties, policy makers, institutions and organisations. There tends to be an alignment between dominant narratives at different levels as there is a continuous process of interaction between top-driven (e.g. government) and bottom-up (people's experiences) narratives. Organisations and projects are expected to respond to the global dominant narratives through their practices and actions. For example, in response to the narrative about sustainability new job roles have been created, e.g. sustainability consultants and other related professional roles. In response to the global narrative of health and wellbeing, organisations tend to organise wellbeing and yoga sessions for their employees. There are new educational programmes and discourses in response to the global narratives. Organisations and projects re-invent and re-label themselves to respond to the global narratives. Hence, narratives are powerful mechanisms through which the world is shaped and influence people. They are often aligned and inter-connected at different levels and contexts.

Narratives can also be temporal as they connect the past, the present and the future (Winch & Sergeeva, 2021). We can distinguish between retrospective narratives, ongoing narratives and prospective narratives. Retrospective narratives are recounted retrospectively after the event has happened, ongoing narrative are narratives about lived experience, and prospective or future-oriented narrative are narratives about future plans, vision and forecasting. All narratives create history and impact our future. Often prospective narratives repeat the past retrospective narratives. This is especially evident in the fashion industry, when new fashion trends are in essence repetition and also re-invention of the old trends.

'Ante-narratives' are what come before a coherent and persuasive narrative, and these are not yet fully formed. Ante-narrative is a gambler's bet that can disrupt and transform narrative thought. In a process of organising both ante-narratives and formed and formalised narration play important roles (Boje, 2008). The future-oriented narratives are visioning or projecting narratives. The future is inherently unknowable, yet projecting involves trying to shape that unknowable future in alignment with the past and present. Leaders may be projecting their personal vision in the case of entrepreneurs such as Elon Musk, who projected his vision as "always to electrify everything with lots of cheap solar power". Musk's passion for cars and space, and, above all, for innovation, led to the creation of Tesla and SpaceX. In the long run, Elon Mask's vision of widespread engagement, at a neurological level, between people and machines, where the human being can form a symbiosis with artificial intelligence.

There are always counter-narratives to the dominant narrative and ongoing interactions between them (Ninan & Sergeeva, 2021). Archival data, such as newspapers and social media messages, can be used as naturalistic sources to help us learn about counter-narratives articulated by external stakeholders and how internal teams deal with these counter-narratives and adapt the narrative in response. This dynamic particularly affects megaprojects due to their significant spatial impact generating conflicting interests between stakeholders. The promoters of a megaproject are interested in supporting its completion, while the protesters are interested in derailing the megaproject. For example, Extinction Rebellion, a global protest movement, is fully opposed nationally and locally to High Speed 2 (HS2) in the UK. Alongside Stop HS2, Extinction Rebellion organised a walk of 200 km along the proposed railway line in June 2020. Project narratives by promoters and protesters will be discussed in greater detail in Chapter 3, and the HS2 case will be presented in Chapter 5. There is a continuous process of interaction between the promoter and protester narratives as the narrative of the project vision evolves in practice. Narratives need to be managed by people. Protesters' narratives may need to be converted into promoters' narratives. Old narratives may need to be re-crafted and become new narratives. In organising, there is a continuous process of re-crafting and renewing narratives in ever-changing world circumstances.

1.6 Past, present and future directions

In this book we discuss the past, present and future of narratives in megaprojects. In this chapter we have defined narratives as well as defining some of its key distinguishing characteristics. We highlight how narratives are coherent as they bring disparate experiences, events and situations together, are performative as they shape and change the future, are strategic as they can create, maintain or disrupt institutions, and are promotional as they are consistently communicated to brand an image of the organisation. Narratives can also exist in written or symbolic/visual forms such as the outlining of corporate objectives in CEO statements, word of mouth circulated among stakeholders, academic publications discussed in seminars, or daily news media articles covering significant events. Narratives can also have identity and sensemaking implications, which we discuss in detail throughout this book. There are also different types of narratives according to their levels, their temporality, and as ante-narratives and counter-narratives.

In Chapter 2, we discuss the importance of narratives in megaprojects. We start by describing what megaprojects are and why making an identity is critical for megaprojects. We highlight the relation between identity and sensemaking and also consider why narratives are required for sensemaking in megaprojects. We then describe how counter-narratives are also present in megaprojects along with narratives and the role of dominant narrative in creating an identity which is essential for the success of megaprojects.

In Chapter 3, we describe the different stakeholders in megaprojects. We highlight promoters as stakeholders interested in supporting the completion of a megaproject, such as government agencies, contractors and lobby groups. We highlight protesters as stakeholders interested in derailing the megaproject such as people affected by the project, activists and non-governmental organisations. The conflict between promoters and protesters are common in megaprojects and we describe how narratives are key to resolving these conflicts. We describe different promoter and protester narratives involved in the need for the project, the stakeholder consultation process, environmental sustainability and excessive noise. We also highlight the interaction between narratives with setting up a narrative, setting up a counter-narrative, and countering the counter-narrative. There is a discussion of the different ways in which the narrative can be contested such as rejecting, delaying and accepting.

In Chapter 4, we describe the different instruments used to craft a narrative. Labels are words or phrases assigned to someone that is instrumental in forming their identity and labels such as 'largest', 'sustainable', 'efficient', etc., are frequently exercised in megaproject settings. Stories are personalised, entertaining and emotional in nature and is instrumental in creating a shared vision amongst organisational members. Comparisons involve comparing oneself with another and are influential in how people make sense of information. We highlight how the narrative instruments of stories, labels and comparisons serve different functions in project organising. While labels help in creating an identity for the project, stories help in creating a shared vision of the project, and comparisons help in enhancing the perception of justice for the community.

In Chapter 5, we describe the different narrative processes to mobilise a narrative. Repeating involves the retelling of narratives in organisations and serves important functions such as control/resistance, integration/differentiation and stability/change within organisations. Endorsing involves getting support from people occupying prominent status in the society and results in trust transference to the megaproject. Humourising involves the use of different forms of comic engagements to make narrative circulate more and reach wider parts of the society. Actioning involves moving narratives from textual/spoken/visual form to an action form and are instrumental in reinforcing the provisional understanding generated through narratives. Thus, in this chapter we highlight how along with repeating, other processes such as endorsing, humourising and actioning can help stabilise the narrative. We also note that stories, labels, and comparisons were individually or together repeated, endorsed, made attractive, and actioned for building the narrative.

In Chapter 6, we record the importance of narrating and storytelling. Whilst we distinguish between narrating and storytelling, we also show their connection.

We discuss a process of interaction between narrating and storytelling in organisational life and its implications for practices of megaprojects. We highlight that project leaders need both narrating and storytelling throughout a project lifecycle. We also discuss formal and informal roles and identities as important parts of narrating and storytelling.

In Chapter 7, we consider the motivations and sense of pride in megaprojects. We discuss different motivations in megaprojects such as pride in the megaproject, the importance of the megaproject, an overemphasis on time, and the acceptance of public inconvenience. We highlight how these motivations are dependent on narratives of instilling pride, shaping identity, creating urgency and working through hardships. We also record how narratives are important for unifying both internal and external stakeholders, necessary for successful completion of a megaproject.

In Chapter 8, we discuss what makes a project narrative and focus on narratives of the future. We outline key features of project narratives. We project our vision about project narratives of future. Narratives of future provide directions and goals for work in projects. Thinking about future creates ideas and new opportunities. We provide future research directions in project narrative work. We suggest some research methodologies and methods to be used.

Throughout this book we also discuss multiple case studies of narratives in megaprojects from across the globe. At the end of this chapter, we discuss the Sochi Olympic Stadium in the Russian Federation to highlight the importance of narratives in megaprojects. In Chapter 2, we discuss the Thames Tideway project, an underground sewer system, in the United Kingdom to highlight the different narratives surrounding the project, such as project identity and project image narrative, and the narrative of innovation. In Chapter 3, we discuss the Melbourne East West Link highway project in Australia to show how conflicts between the promoters' the and protesters' narratives can lead to the cancellation of a project. In Chapter 4, we discuss the Dakota Access Pipeline in the USA to highlight how the applying of labels to the protesters can be instrumental in the creation of a narrative discrediting them. In Chapter 5 we discuss the High Speed 2 rail project in the UK to record the different instruments and processes in mobilising a narrative. In Chapter 6 we discuss the Crimean Bridge in the Russian Federation to highlight the role of storytelling in shaping narratives in megaprojects. In Chapter 7 we discuss the Chennai metro rail in India to show how narratives can create a sense of pride in both internal and external stakeholders. Finally, in Chapter 8 we present the Eden Project case, an eco-tourist project in the United Kingdom, to describe the role of narratives of future, and how it has been expanded worldwide. The case studies and their narrative perspective is summarised in Table 1.1.

The case studies cover both the successful and the unsuccessful use of narratives, highlighting that narratives in megaprojects are important and that the extent to which they prevail can have an effect on whether or not they are constructed.

TABLE 1.1 Case studies and their narrative perspectives

Chapter	Name of project	Type of project	Country of project	Narrative perspective
1	Sochi Olympic Stadium	Stadium	Russian Federation	Importance of narratives
2	Thames Tideway	Sewer system	United Kingdom	Narrative of project identity and image Narratives of innovation
3	Melbourne East West Link	Highway	Australia	Conflict between promoter and protester narratives
4	Dakota Access Pipeline	Pipeline	United States of America	Labels and narratives to discredit protesters
5	High Speed 2	Railway	United Kingdom	Different instruments and processes in mobilizing a narrative
6	Crimean Bridge	Bridge	Russian Federation	Role of storytelling in shaping narratives
7	Chennai metro rail	Metrorail	India	Narratives for sense of pride
8	Eden Project	Tourism	United Kingdom	Narratives of the future

1.7 Conclusion

In this chapter we have defined narratives as well as outlining some of their key distinguishing characteristics. We highlighted how narratives are coherent as they bring together disparate experiences, events and situations. Narratives are also performative in nature as they shape and change the future. They are strategic as they can create, maintain or disrupt institutions. Finally, narratives are promotional as they are communicated consistently to brand an image of the organisation. Narratives can also have identity and sensemaking implications, something which we discuss in detail throughout this book. There are also different types of narratives according to their levels, their temporality, and their status as ante-narratives and counter-narratives. In the next chapter, we focus on megaproject narratives and provide more practical examples.

CASE STUDY 1: SOCHI OLYMPIC STADIUM, RUSSIAN FEDERATION

The compelling narrative about Fisht Sochi Olympic Stadium megaproject

Fisht Olympic Stadium is an outdoor stadium in Sochi, Russian Federation. Located in Sochi Olympic Park and named after Mount Fisht, the 40,000-capacity stadium was constructed for the 2014 Winter Olympics and Paralympics, where it served as the venue for their opening and closing ceremonies. The stadium was originally built as an enclosed facility; it was re-opened in 2016 as an open-air football stadium, to host matches as part of the 2017 FIFA Confederations Cup and the 2018 FIFA World Cup, when it was known simply as Fisht Stadium. This case study tells the story about how different narratives have been constructed of Sochi Olympic infrastructure.

The architectural concept of Fisht Stadium was developed by a consortium consisting of: GUP MNIIIP "Mosproekt-4" (Russia), the architectural bureau Populous (Great Britain), project management specialist Botta Management Group (Switzerland), and construction contractor company CJSC Association Ingeocom. All participants in the consortium created specifically for participation in the tender for the construction of the main Olympic stadium have extensive experience in implementing projects of this level, including Olympic ones.

One of the most important issues that experts paid special attention to was the post-Olympic use of the structure. The central stadium was designed not only taking into account the possibility of holding events related to the 2014 Olympics, but also taking into account the subsequent operation: if necessary, the central stadium can be transformed with a change in the number of seats and functional zoning. The solutions laid down in the projects provide a high level of comfort for participants and guests of the Olympics, and also take into account the needs of people with disabilities.

The Central Olympic Stadium was originally designed as one of the most ambitious buildings of the Olympic Games, their architectural symbol. Its composition is absolutely unique to Russia. The translucent polycarbonate roof gives its appearance a resemblance to a snowy peak that harmoniously merges into the panorama of the Caucasus Mountains. In accordance with the concept of the 2014 Olympic Games, it was assigned one of the dominant roles. The presented construction is not just for sports facilities; it is, in essence, theatrical stage where sporting events take place, and spectators must certainly feel the special atmosphere of the Olympics. Figure 1.2 shows a render of the stadium and Figure 1.3 shows a picture of the stadium.

FIGURE 1.2 Rendering of the Sochi Olympic stadium

FIGURE 1.3 Completed Sochi Olympic stadium

The narrative about environmental sustainability

During the preparation and conduct of the Winter Olympic and Paralympic Games, the city of Sochi offered its participants and guests a unique natural environment. The Winter Olympics have never been held in such a diverse natural environment. The games venues were located within natural landscapes in direct contact with the Sochi National Park and other specially protected areas. To preserve and improve the quality of the environment during the preparation of the Olympics, a comprehensive plan was developed, including a number of environmental measures that ensure the implementation of environmental goals and commitments.

In 2009 the Sochi-2014 Environmental Strategy was approved for organising a "green" Winter Olympics in the city. The organisers were set a number of ambitious goals and objectives:

- to protect the unique nature of the Sochi region, including the protected natural complexes of the Sochi National Park and the Caucasian State Natural Biosphere Reserve;
- to improve the environment of the city of Sochi through the large-scale development of the regional infrastructure; and
- to unite the efforts and resources of all interested parties for the best organisation of the 2014 Winter Olympic Games in Sochi and the formation of a positive heritage for future generations.

The structure of the Sochi-2014 Environmental Strategy was composed of four strategic directions, each of which was given a symbolic name: "Games in harmony with nature", "Games without climate change", "Games without waste", and "Enlightenment Games".

As a result of the Olympic project, the city of Sochi received a significant tangible and intangible ecological heritage. The material heritage can be attributed, first of all, to the provision of environmentally effective sports facilities that form the first national centre of winter sports; the Olympic venues were certified according to the international BREEAM standard and according to the Russian Green Standards system. In addition, the city received a regional infrastructure offering environmentally-efficient solutions necessary for the development of Sochi in the fields of transport, energy generation and supply, water supply, sanitation and waste management.

The intangible environmental heritage includes the "green standards" of construction, focused on energy and environmental efficiency; an environmental management system implemented at Olympic venues based on the international standard ISO 14001: 2009 and an increased environmental awareness of residents of the Sochi region. Thus, the city of Sochi received a new infrastructure and significant environmental heritage. However, not all of the goals stated in the above programme documents have led to the expected results. The environmental impacts of the Olympic project in the two main areas of the Sochi-2014 Environmental Strategy turned out to be the most problematic: "Games in harmony with nature" and "Games without waste".

Sochi's organisers have failed with regard to their green promises, says Suren Gazaryan, a zoologist and member of the environmental campaign group Environmental Watch of the North Caucasus (EWNC). Gazaryan explained that the construction process for the games has been hugely damaging for the region. He and the ENWC have documented evidence of illegal waste dumping, construction that has blocked the migration routes of animals such as the brown bear, limited access to drinking water for locals and a generally decreased quality of life for many in the city of Sochi. "The most dangerous and important part of the damage is the biodiversity lost in the area", says Gazaryan. "Parts of the national park have been completely destroyed. This area was the most diverse in terms of plant and animal life in Russia." There is also the added danger of increased landslides, mudflows and building collapses as a result of poor construction and hazardous waste dumping practices, says Gazaryan.

Simon Lewis, who runs Team Planet, a UK-based consultancy on sustainability in sport, states that the Sochi organisers already had their work cut out for them. Hosting a Winter Games is often more challenging from an environmental perspective than hosting the Summer Games: "If you look at the environmental footprint of hosting a Games – including things like travel, construction and hospitality – doing that halfway up a mountain in what is often a delicate and pristine environmental habitat is going to be difficult", he says. The International Olympic Committee (IOC) and the United Nations Environment Program (UNEP) worked with organisers to help mitigate some aspects of the construction, including relocating some sporting venues away from the borders of the UNESCO World Heritage Site. Despite this, says Lewis, "Sochi should never have happened in that location. It was a poor decision by IOC members based on poor information."

Gazaryan was cynical about whether or not the unresolved environmental concerns would receive much attention when the spectacle of the games gets underway. Like Lewis, he believed that the IOC needed to reconsider how it ensures sustainability standards are met in future. Given Russia's hopes of making Sochi a global ski destination after the Games – which would open up a sensitive national park region to increased tourist traffic – it seems unlikely that its environmental legacy is one the IOC will be shouting too loudly about in the years to come.

The narrative about the cost overrun of Fisht Stadium

How has Fisht Stadium risen in price from 1.6 billion rubles to 17 billion and become twice as expensive as any other stadium in the world?

According to the main architect of Fisht Stadium:

> If the terms of reference for the design of the central Olympic stadium, now Fisht, were written about what the opening and closing ceremonies would look like, all these costs could have been avoided to a large extent. But, unfortunately, as often happens with us, the need for an indoor hall instead of an open stadium was revealed only a year before the end of construction. On the go, I had to change all the technical solutions which leads to a rise in price two times. We always have this problem, not only at the central stadium, but also at other sites of the Olympic Games. Nevertheless, everyone coped with the tasks and held the Olympics.

He further discusses that

> in more than ninety percent of the cases, the opening and closing ceremonies are held in open stadiums, converted football or athletic, where there are treadmills and all areas for athletics around the football field. Until the last two Olympics, everyone was happy with that. Although everyone understands that the ceremony in the concert hall, with light, sound, decorations and all the technical capabilities, is, of course, more spectacular. After long discussions of the options, they approved such

a decision, angry and reliable – to make a roof over the entire football field. But not always the most understandable solution – the most economical. When the construction of two visitors over the western and eastern stands had already begun, along the way, it was necessary to make changes to the design and re-pass the examination, working documentation, which is the most serious in the course of construction work. This affected the timing of work, and their cost. And it affects so far, because our system of concluding government contracts involves penalties for failure to meet deadlines. And our general contractor, who led the construction of the central stadium, if I am not mistaken, is still suing the heirs of the state corporation Olympstroy.

His concluding statement about the cost of the project:

The figure was determined incorrectly by a poor-quality concept, a poor-quality first project, which was made by an unprofessional organisation, which before that was engaged in the design of agricultural facilities, offices, large shops. This is a system for holding a tender for design work: it is not the one who professionally can do it wins, but the one who offered the cost and terms that satisfied the customer. We will not talk about corruption components, but about our tenders for design work. To win, you don't need to come up with any design solution, you need to submit two papers: the cost of work and the deadlines plus some documents, including bank guarantees. The second problem is that at the project stage no one knows how the competition will be held. That is, the technical task is done by the customer, Olimpstroy, where there were no specialists in sports facilities or ceremonies.

Exercise

From the above case study, please answer the following questions:

1) Why is a compelling narrative needed for a megaproject such as Fisht Sochi Olympic Stadium?
2) To what extent is a narrative about the purpose of Fisht Sochi Olympic Stadium megaproject connected to the global narrative about sustainable development and the challenges of climate change?
3) Identify counter-narratives of Fisht Sochi Olympic Stadium megaproject and by whom these are created.
4) What can we learn from a narrative of Fisht Sochi Olympic Stadium megaproject that could inform future megaprojects?

Sources

Designerdreamhomes. (2014). The unique project of the famous Fisht Stadium in Sochi. https://designerdreamhomes.ru/sovremennyy-stadion-fisht-ot-kompanii-populous/ (Accessed on 19 August 2022).

Makarov, I. (2015). Sochi a year later: "It was possible to design the stadium without reconstruction, but not today, but eight years ago". https://republic.ru/posts/l/1212767 (Accessed on 19 August 2022).

Müller, M. (2014). (Im-)mobile policies: Why sustainability went wrong in the 2014 Olympics in Sochi. *European Urban and Regional Studies*, 22(2), 143–160.

Paramaguru, K. (2014). The Not So Sustainable Sochi Winter Olympics. https://time.com/2828/sochi-winter-olympics-environmental-damage/ (Accessed on 19 August 2022).

Zakharova, N. (2011). The final design of the Central Stadium Sochi-2014. http://arch-sochi.ru/2011/08/okonchatelnyiy-proekt-tsentralnogo-stadiona-sochi-2014/ (Accessed on 19 August 2022).

Zakharova, N. (2016). Olympic project in Sochi: Environmental aspects. http://arch-sochi.ru/2016/01/olimpiyskiy-proekt-v-sochi-ekologicheskie-aspektyi/ (Accessed on 19 August 2022).

References

Alvesson, M., Ashcraft, K. L., & Thomas, R. (2008). Identity matters: Reflection on the construction of identity scholarship in organization studies. *Organization*, 15(1), 5–28.

Boje, D. M. (2001). *Narrative methods for organizational and communication research.* London: Sage.

Boje, D. M. (2008). *Storytelling organizations.* London: Sage.

Brown, A. D. (2006). A narrative approach to collective identities. *Journal of Management Studies*, 43(4), 731–753.

Bucholtz, M., & Hall, K. (2005). Identity and interaction: A sociocultural linguistic approach. *Discourse Studies*, 7(4–5), 585–614.

Cunliffe, A., & Coupland, C. (2011). From hero to villain to hero: Making experience sensible through embodies narrative sensemaking. *Human Relations*, 65(1), 63–88.

Currie, G., & Brown, A. D. (2003). A narratological approach to understanding processes of organizing in a UK hospital. *Human Relations*, 56(5), 563–586.

Czarniawska, B. (1998). *A narrative approach in organization studies.* Thousand Oaks, CA: Sage.

Dailey, S. L., & Browning, L. (2014). Retelling stories in organizations: Understanding the functions of narrative repetition. *Academy of Management Review*, 39(1), 22–43.

Frandsen, S., Kuhn, T., & Lundholt, W. (2017). *Counter-narratives and organization.* London: Routledge.

Gioia, D. A., Schultz, M., & Corley, K. G. (2000). Organizational identity, image, and adaptive instability. *Academy of Management Review*, 25(1), 63–81.

Gregory, J. R. (2004). *The best of branding: Best practices in corporate branding.* New York: McGraw-Hill.

Hansen, P. H. (2012). Business history: A cultural and narrative approach. *Business History Review*, 86(4), 693–717.

Hjorth, D., & Steyaert, C. (2004). *Narrative and discursive approaches in entrepreneurship.* Northampton, MA: Elgar.

Humphreys, M., & Brown, A. D. (2002). Narratives of organizational identity and identification: A case of hegemony and resistance. *Organization Studies*, 23(3), 421–447.

Kalkan, Ü., Altınay Aksal, F., Altınay Gazi, Z., Atasoy, R., & Dağlı, G. (2020). The relationship between school administrators' leadership styles, school culture, and organizational image. *Sage Open*, 10(1), 2158244020902081.

Kärreman, D., & Alvesson, M. (2004). Cages in tandem: Management control, social identity, and identification in knowledge-intense firm. *Organization*, 11(1), 149–175.

Lawrence, T. B., & Suddaby, R. (2006). Institutions and institutional work. In S. R. Clegg, C. Hardy, W. R. Nord, & T. Lawrence (eds.), *Handbook of organization studies* (pp. 215–254). Thousand Oaks, CA: Sage.

Martens, M. L., Jennings, J. E., & Jennings, P. D. (2007). Do the stories they tell get them the money they need? The role of entrepreneurial narratives in resource acquisition. *Academy of Management Journal*, 50(5), 1107–1132.

Mathur, S., Ninan, J., Vuorinen, L., Ke, Y., & Sankaran, S. (2021). An exploratory study of the use of social media to assess benefits realization in transport infrastructure projects. *Project Leadership and Society*, 2, 1–10

Ninan, J., Clegg, S., & Mahalingam, A. (2019). Branding and governmentality for infrastructure megaprojects: The role of social media. *International Journal of Project Management*, 37(1), 59–72.

Ninan, J., & Sergeeva, N. (2021). Labyrinth of labels: Narrative constructions of promoters and protesters in megaprojects. *International Journal of Project Management*, 39(5), 496–506.

Ninan, J., Sergeeva, N., & Winch, G. (2022). Narrative shapes innovation: A study on multiple innovations in the UK construction industry. *Construction Management and Economics*, 1–19. https://doi.org/10.1080/01446193.2022.2037144

Pentland, B. T. (1999). Building process theory with narrative: From description to explanation. *Academy of Management Review*, 24(4), 711–724.

Pentland, B. T., & Feldman, M. S. (2007). Narrative networks: Patterns of technology and organization. *Organization Science*, 18(5), 781–795.

Polkinghorne, D. E. (1991). Narrative and self-concept. *Journal of Narrative and Life History*, 1(2–3), 135–153.

Rhodes, C., & Brown, A. D. (2005). Narrative, organizations and research. *International Journal of Management Reviews*, 7(3), 167–188.

Sims, D. (2003). Between the millstones: A narrative account of the vulnerability of middle managers' storying. *Human Relations*, 56(10), 1195–1211.

Sergeeva, N., & Winch, G. M. (2021). Project narratives that potentially perform and change the future. *Project Management Journal*, 52(3), 264–277.

Sonenshein, S. (2010). We're changing – Or are we? Untangling the role of progressive, regressive, and stability narratives during strategic change implementation. *Academy Management Journal*, 53(3), 477–512.

Stets, J. E., & Burke, P. J. (2000). Identity theory and social identity theory. *Social Psychology Quarterly*, 63(3), 224–237.

Thomas, R., & Davies, A. (2005). Theorising the micro-politics of resistance: Discourses of change and professional identities in the UK public services. *Organization Studies*, 26(5), 683–706.

Tomkins, L., & Eatough, V. (2013). The feel of experience: Phenomenological ideas for organizational research. *Qualitative Research in Organizations and Management: An International Journal*, 8(3), 258–275.

Trible, P. (1984). *Texts of terror: Literary-feminist readings of biblical narratives*. Philadelphia, PA: Fortress Press.

Vaara, E., Sonenshein, S., & Boje, D. (2016). Narratives as sources of stability and change in organizations: Approaches and directions for future research. *The Academy of Management Annals*, 10(1), 495–560.

Winch, G. M., & Sergeeva, N. (2021). Temporal structuring in project organizing. *International Journal of Project Management*, 40(1), 40–51.

2

MEGAPROJECTS AND NARRATIVES

2.1 The context of megaprojects

Megaprojects are temporary organisations characterised by a large-scale investment commitment, complexity and long-lasting impact on the economy, the environment and society. "Megaprojects are large-scale, complex ventures that typically cost $1 billion or more, take many years to develop and build, involve multiple public and private stakeholders, are transformational, and impact millions of people" (Flyvbjerg, 2017, p. 2). Examples of megaprojects are airports, city re-building, railways, highways, stadiums, power plants, bridges, tunnels, dams etc. These refer not only to construction projects but also to decommissioning projects. They are flexible and diverse in their purposes and business models. Megaprojects attract a lot of attention from the public because of substantial long-lasting impacts on economy, environment and communities. These types of projects create value for the economy, society and the environment by creating new assets, job opportunities and services. At the same time, megaprojects are displacing people (e.g., those people who live in the local area where the megaproject is build may need to re-locate) and disrupt local environment (e.g., at the time of constructing a megaproject, the pollution may be higher than usual). Thus, megaprojects cause social, economic and political disruptions in an area.

The rationale on which megaprojects are built is collective benefits. For example, a highways megaproject provides value in the terms of the number of people it can transport, the number of goods it can transport, the ability to provide livelihood to a lot of people, attracting new shopping malls to the area, and the provision of additional services around the highways. However, megaprojects have been criticised for their top-down planning processes, negative effects and consequences on communities. Megaprojects often advantage one group of people whilst disadvantaging another. The main reasons policy makers are attracted to megaprojects

DOI: 10.1201/9781003248378-2

are: a) Technological sublime: The rapture that engineers and technologists get from building large and innovative megaprojects, pushing the boundaries for what technology can do; b) Political sublime: The rapture politicians get from building monuments to themselves and for their causes; c) Economic sublime: The delight business people and trade unions receive from the profits and jobs created by megaprojects; and d) Aesthetic sublime: The pleasure designers and people who appreciate good design get from building, using and looking at something very large that is also 'ironically beautiful' (Flyvbjerg, 2014, 2017). Another related sublime is community pride (Söderlund et al., 2017), which is the ability of megaprojects to provide a sense of accomplishment for the community wherein they tell stories about the big things they accomplished and making sure. Because of these sublimes, more and more projects are being delivered as megaprojects, and the general size of megaprojects is becoming larger.

However, even though the number and size of these projects increase worldwide, these projects are shrouded in inefficiencies and failures. Flyvbjerg (2003) studied 258 megaprojects in 20 countries, and conclude that 90 per cent of them fail to deliver on their planned objectives, time, cost or promises to stakeholders. Megaprojects often go over budget and over time. During the shaping phase, concerns over cost overruns are often expressed by the critics of megaprojects, since this is a frequent outcome. Other indicators of failure in these projects include negative environmental effects such as noise, pollution, landscape erosion, etc. (Priemus et al., 2008). These projects have social impacts, along with the environmental and economic impacts, as large numbers of people are inconvenienced and sometimes displaced to enable the construction of the megaproject. Due to the large-scale nature of these projects, any failure in managing them can affect the economy, a variety of stakeholders and not bring about the socioeconomic development envisioned. The US$1 billion cost overrun of the 2004 Athens Olympics, is a case in point, since it almost single-handedly contributed to the economic crisis in Greece (Flyvbjerg, 2007). We have also included case studies of the Melbourne East West Link, which was an economic failure, in Chapter 3 and the Dakota Access Pipeline, which was a social failure, in Chapter 4. Thus, megaprojects are a high-risk venture with both a higher probability of failure and a higher impact on failure. Hence, megaprojects are a mega-challenge for management and project management scholars worldwide are trying to understand megaproject practice and how to construct them efficiently.

Megaprojects are about people and teams. Political leadership is required to secure resources, generate public support, mollify critics and manage conflicting interests throughout a project lifecycle. Other challenges faces are laws and regulations that empower community groups, contested information, a high level of uncertainty and complexity, avoiding any impact on neighbourhoods and the environment. Managing and leading megaprojects is about people who make decisions that enable actions (Morris & Pinto, 2010; Volker, 2019). The actions, practices and behaviours of the megaproject participants are to be explained by their motivations and human needs. After all, decision-making is a social behaviour even when nobody else is present, as the decision-maker can anticipate how others will react and factor it into their decision. Constructs such as megaproject mission and vision, sustainability, innovation, identities and image at the individual and collective levels are social constructions often created by discourses and narratives. Hence, megaprojects can be better understood as continuously reconstructed entities with narratives driving experiences through the dramatizing trajectories of practice.

2.2 Identity in megaprojects

Identity is a social and cultural phenomenon that encompasses macro-level categories, temporary and interactionally specific stances, and cultural positions (Bucholtz & Hall, 2005). The concept of identity is central to organising as it helps in the development of shared interests and goals (Hietajärvi & Aaltonen, 2018). It affects how an individual, organisation, industry or nation interprets and responds to institutional forces (Kodeih & Greenwood, 2014). Due to this, organisational identity is referred to as a 'close-to-the-bone' concept by Gioia et al. (2013) as they call for more empirical work to better understand its origins and transitions. The development of an identity in an organisation enables stakeholders to articulate shared interests and commit to preferred outcomes, and thereby the necessary action (Ashforth et al., 2008).

Identities at the individual and the collective levels are social constructions (Gioia et al., 2000), often shaped by language through personal and shared narratives (Humphreys & Brown, 2002). Marketing plays an important role in navigating a complex and turbulent environment by creating an identity (Christensen, 1995). The most obvious way to constitute identities is through the overt introduction of referential identity categories, such as narratives, into discourses (Bucholtz & Hall, 2005).

Theorists claim that organisations selectively assign negatively-valued social identities such as 'poor credit risk' to those whom they disfavour (Ashforth & Humphrey, 1997) and, conversely, they assign positively-valued identities such as 'fair employer' to those whom they prefer (Albert & Whetten, 1985). By projecting a narrative, an organisation stakes a claim to a status or identity similar to the category it claimed (Pontikes, 2018) that might be difficult to establish by other means (Ashforth & Humphrey, 1997). Thus, narratives can be considered as a tag for defining an organisation's identity (Ashforth & Humphrey, 1997).

Within the context of megaprojects, organisational identity stimulates the diverse stakeholders to commit themselves to the megaproject (Sergeeva & Ninan, 2022; Van Marrewijk, 2007). This commitment is based on the membership to the group combined with the emotional value that is attributed to this membership (Veenswijk et al., 2010). Since identity is created through narratives, megaprojects are prone to different narratives, from both the promoters' as well as the protesters' group. As Sergeeva (2017) notes, narratives that shape megaproject identities are discursive and are constructed, re-constructed and maintained through multiple processes of contestations. Of these, the promoters' and protesters' narratives are significant as they strive to shape the identity of different parts of the project.

The case of the Thames Tideway megaproject in the United Kingdom, outlined in this chapter, shows how project identity has been created, and the crucial role played by the Chief Executive Officer in identity construction.

Building on organisational identity theory (Schultz & Hernes, 2013), Sergeeva and Roehrich (2018) focus on narratives and their associated meanings in megaprojects. They find that megaprojects socially construct their identities as 'learning organisations' via spoken, symbolic and written forms of narratives: sharing stories and videos via digital platforms and also by writing reports and blogs. There is a need to understand the processes through which these often-conflicting actors create and transform their institutional contexts (Esposito et al., 2021). It is in this context that a megaproject narrative, which portrays different parameters of the megaproject including the mission and vision of the megaproject, is crucial for navigating the complexity, uncertainty and conflictive nature of the situation and bringing about change in the society.

2.3 Sensemaking and identity

People in identity crisis use sensemaking to extract cues and make plausible sense while enacting order into ongoing circumstances (Weick et al., 2005). Multiple literature notes that language and the variety of communication instruments can be treated as a building block of sensemaking (Boyce, 1995; Watson & Bargiela-Chiappini, 1998). After all, according to Weick (1995), Weick et al. (2005), sensemaking is about noticing, bracketing and labelling. Activities may be assigned to categories in ways that predispose practitioners to find common sense. To find the common sense, a narrative ignores differences amongst actors and deploys cognitive representations. Weick et al. (2005) articulate that narrative could fail to capture the dynamics of what is happening as it follows after the completed act. Corley and Gioia (2004) and Gioia et al. (2000) reinforce that narratives are used to describe "who they are" and "what they are doing" and, even though it may be sustained over time, the meanings and interpretations associated with them may change. As the process unfolds over time, narratives can evolve. Weick (1995, p. 31) suggested that when individuals enact, they: "Undertake undefined space, time, and action, and draw lines, establish categories, and coin new labels that create new features of the environment that did not exist."

Language is not just a mirror of reality; it is also a force shaping how processes occur and events emerge (Phillips & Oswick, 2012). Good communication is widely acknowledged as essential for project success (Reed & Knight, 2010); however, with sensemaking language is seen as representing and constructing project events (Havermans et al., 2015). Language can also be considered as a resource to manage tensions, as noted by Gil (2010) in his analysis of the verbal accounts produced by senior project management. A focus on linguistic practices is important for exploring everyday project realities (Packendorff, 1995). Projects are seen as co-constructed in everyday communicative interactions (Lindgren & Packendorff, 2007) and research needs to consider how what people do in projects is embedded in, and shaped by, social processes of interaction and communication (Cicmil et al., 2006). Of the various linguistic practices, a study on narratives and its implications can help us understand projects better and extend the perception of the use of language which is currently limited to communication.

Regarding agencies involved in the narrative process, Pontikes (2012) records that narratives are not a product of a single actor; rather, they are a product of competing claims from multiple audiences. Different parties in an organisational setting campaign and compete to shape meanings of and in the organisation, to gain acceptance for a preferred account, or to subvert the status quo. Accordingly, there are tussles and tensions of sensemaking (Maitlis & Christianson, 2014). Narrative contestations occur when two or more stakeholders attempt to define divergent realities for a given audience (Ashforth & Humphrey, 1997). To navigate these contestations, direction, in the form of top-down sensemaking, is important (Ancona, 2011) as leadership is the 'management of meaning' (Smircich & Morgan, 1982). This top-down approach is also referred to as guided sensemaking (Maitlis & Lawrence, 2007) or sensegiving (Gioia & Chittipeddi, 1991).

2.4 Narratives in megaprojects

Multiple stakeholders have to work together to successfully complete a project. Internal stakeholders can be managed by systems integration through innovative contracts (Davies et al., 2009), decomposing the project into manageable entities (Davies & Mackenzie, 2014), improving communication across these organisations (Roehrich & Lewis, 2014) and collective decision-making processes (Gil & Pinto, 2018). However, external stakeholders cannot be governed by such contractual instruments. In such situations where different stakeholders have different objectives and are difficult to govern, there is a need to achieve strategic convergence among these conflicting objectives (Denis et al., 2007).

A common narrative is essential to drive the project among these conflicting goals. A common project narrative is a narrative about a project, what it aims to achieve mobilising human and financial resources. The common megaproject narrative helps to have a shared understanding and vision about team members to work together towards project deliverables and outcomes (Sergeeva & Winch, 2021). It helps to create a shared purpose for a project team. The common, dominant project narrative needs to be crafted and maintained throughout a project lifecycle among other conflicting objectives and counter-narratives. The dominant project narrative overcomes resistance to change and counter-narratives. After all, narratives help create common identities by bringing plausibility and coherence to disparate experiences (Grayson, 1997; Humphreys & Brown, 2002; Vaara & Tienari, 2011). As noted by Sturup (2009), narratives have significant importance in the context of megaprojects which cause significant environmental, social and political disruptions in its local environment. By structuring a message as a narrative, it becomes more persuasive as people become absorbed in a story than an analytical illustration of a product's features which distracts people's attention (Escalas, 2007). Narratives of purpose, relevance and scope act as 'gatekeeper' in terms of inclusion and exclusion of meaning, and thereby influence the meaning production within the community (Veenswijk et al., 2010). Project narratives are crafted for defining the project mission, getting approvals from sponsors, and convincing internal and external stakeholders. A project mission narrative is a compelling narrative about why a project is needed. A project scope narrative is a compelling narrative how a project mission is going to be achieved. Megaprojects require favourable narratives as such narratives can build strong brand attitudes and brand-loyal behaviours (Grayson, 1997; Ninan et al., 2020). As Olander and Landin (2008) note, it is important to brand the project with good reputation and media image right from its start and hence building a narrative up front is essential for the successful delivery of projects from an external stakeholder management perspective. Project narratives have important implications for achieving project outputs (e.g. deliver on time and on budget) and outcomes (e.g. create value for customers). Below are two examples of project mission narratives.

The project mission narrative of High Speed 2 Ltd. (HS2) is: "We are building a new high speed railway to better connect people across Britain. As a high performing, innovative organisation, we will deliver value for money by applying the best in worldwide design and construction techniques" (www.hs2.org.uk). HS2 presents its image project narrative as an innovative, high-performing organisation that directly addresses the government's narrative about the need for innovation. It also addresses international narrative through a narrative of applying best techniques worldwide. This case is presented in Chapter 5, where more details about project narratives, and how these are crafted and maintained are discussed.

The project mission narrative of Thames Tideway is

> We want to create a wider legacy for the capital and, in doing so, help realise the vision to reconnect London with the River Thames. The primary task of the project is to provide the infrastructure that London needs to flourish economically and socially for decades to come.
>
> (www.tideway.london)

As evident, the mission project narrative addresses the government's narrative about the need to innovate. It also addresses the national and international narrative about a sustainable future. The case is presented at the end of this chapter where further details are discussed.

Project narratives are communicated in textual (e.g. website, reports, strategies), spoken (e.g. presentations, talks) and visual (e.g. videos, pictures, social media) forms. Project leaders' and their team's beliefs and commitments turn project narratives into commitments and actions. Communicating project narratives in all three forms strengthens and reinforces the message and helps to achieve consistency. Promoting project narratives in social media becomes popular in a form of posts, pictures and short videos (Mathur et al., 2021). This captures attention from the general public, capture their project experience and even store these information digitally. Project image narratives are crafted for purposes of sharing experiences, promoting project stories and helps to win new projects.

Organisational practices, such as achieving strategic convergence, can be understood as symbolic manipulation being combined with persuasive rhetoric, and there is a need for investigating such phenomena in the context of megaprojects (Bresnen & Marshall, 2001; Ninan et al., 2022). Sorsa and Vaara (2020) have longitudinally studied conflicts between proponents and challengers in building a new parking centre in the historic square and constructing new high-end condos in the downtown harbour area in a Nordic city. They note how the use of narratives and rhetoric were instrumental in moving from initial contestation through gradual convergence to increasing agreement between the proponents and challengers. Hence, stakeholder's acceptance towards a project can be achieved through narratives. We argue that in spite of the importance of narratives in achieving strategic convergence of objectives in megaprojects and thereby external stakeholder management, the practice of mobilizing narratives in the context of megaprojects are yet to be explored, a gap this book is set to bridge.

Megaprojects are bespoke, one-off endeavours with a specific end date, but usually a long lifespan throughout where managers and the composition of project teams keep changing. Additionally, in megaprojects different organisational identities and cultures merge together (e.g. owners, suppliers, consultancies, and so forth), which creates a need for a megaproject identity narrative or, in other words, a narrative about project DNA that is created and communicated to internal team members in order to align effort. Megaprojects also attract attention from wide audiences (e.g. social media, external stakeholders, press) which requires the consistent communication of a coherent narrative about the megaproject brand image to external stakeholders. Hence, there is a need for creation of a narrative about the mission, vision and expected benefits and value before a megaproject starts, during its lifespan and beyond. Maintaining these project narratives throughout the project lifecycle is often required to achieve shared understanding among the team members and wider communities.

2.5 Counter-narratives in megaprojects

Narratives, however, are unstable and shift from one equilibrium to another (Todorov, 1971). The equilibrium is a dominant narrative that is generally accepted as a universal truth (Harper, 2009). The stories people say that offer resistance to, either implicitly or explicitly to the dominant narrative are called counter-narratives (Andrews, 2004). In suggesting how else it could be told, counter-narratives expose the construction of the dominant story (Harris et al., 2001). Boje (2001) calls the alternative stories that were not part of the shared vision as 'rebel voices'. These counter-narratives help to document, and perhaps even validate, a 'counter-reality' (Delgado, 1995).

Counter-narratives undermine the shared and explicit narrative (Zilber, 2007) and strive to create a new dominant narrative. Exploring counter-narratives enables us to understand the struggles over meanings, values and identities that take place in organising (Frandsen et al., 2017; Ninan & Sergeeva, 2021). It helps us to capture some of the political, social and cultural complexities and tensions in organising (Sergeeva, 2019). McQuillan (2000) claims that the contest between the narrative and counter-narrative structures the narrative matrix and records counter-narrative as a necessary condition for narrativity. The boundary between dominant and counter-narratives is not fixed, and they are always less stable and unified than they appear (Squire, 2002). With time and effort, counter-narratives can potentially change the dominant narrative (McLean & Syed, 2015). In the process of creating the counter-narrative, individuals reference the dominant narrative and position themselves against, or in contrast to it (Andrews, 2002; McLean, 2015). However, how organisations resist counter-narrative is still not explored.

A megaproject setting offers an avenue to explore the interaction between narratives and counter-narratives due to the shorter time span in contrast to social and cultural changes. Within project settings, Veenswijk and Berendse (2008) argue that narratives are important vehicles through which meanings are negotiated, shared and contested. They highlight that an analysis of project narratives helps understand organisational change processes. Counter- and competing narratives inevitably arise in projects, such as megaprojects, which involve uncertainty, integration and urgency (Boddy & Paton, 2004). Drevin and Dalcher (2011) record the presence of multiple counter-narratives before the coherent post-project narrative of success or failure emerges. As noted by Boddy and Paton (2004), the sources of counter-narratives lie in people's subjectivity while they interpret the distinguishing features of a project and its context. Discussing leaders' narratives, Havermans et al. (2015) highlight how leaders frame the project and dictate whether the project is routine or ground-breaking. They note that when the goals and methods of a project are unclear, the narrative proposed by the leader is likely to be more fluid and negotiable than when project goals and methods are clear. In the context of megaprojects, the promoters and protesters strive to create a narrative of the project either to stabilise or to destabilise the project. Taking the instance of policy measures in megaprojects, Esposito and Terlizzi (2019) record the process through which promoters and protesters of a given policy measure engage in battles over competing narratives. The stakeholders advocate arguments either in favour or against a particular policy in order to influence its final outcome. Adding to the gap in organisation studies, i.e., an exploration on how organisations resist counter-narratives, from a project management perspective understanding the dynamic between narratives and counter-narratives can help projects manage community resistances and deliver projects effectively.

2.6 Narratives throughout a project lifecycle

Different project narratives can be identified depending on the specific project nature (Sergeeva & Winch, 2021; Winch et al., 2022). Figure 2.1 shows different

Project image narrative for external stakeholders	Project identity narratives for team members	Project narratives about value created for society
Narratives about project mission, vision and desired future	Project identity narratives for supply chain	Project narratives about realised outputs and outcomes
Project narratives about expected value	Narratives about how to deliver intended project outputs: on time, on budget and to quality requirements	Narratives about project awards and achievements

$\longrightarrow$ Project lifecycle

Project shaping	Project delivery	Post-project evaluation

FIGURE 2.1   Project narratives throughout a project lifecycle

project narratives in relation to a project lifecycle. At project shaping phase project narratives about mission, vision and desired future are to be crafted, clearly stated, become formalised and then communicated to a public. Stakeholders who initiate a project craft and maintain a dominant project narrative. It is during the earlier stages of a project lifecycle, an image-shaping narrative is articulated with the purpose of projecting the desired future to external stakeholders. The project image narrative tends to be favourable and optimistic. Project counter-narratives crafted by external stakeholders who could be protesters or promoters may be observed during the shaping phase of a project. During the project delivery phase, the project identity narrative is expected to be created and maintained. The project identity narrative is created by the owner project team and shared with the supply team. In a temporary nature of project environment, different cultures and identities from different organisations and people involved in a project merge together, forming a project identity narrative. A project identity narrative is important as it is about a project's DNA. In the process, the project team members and those involved feel part of a project work and its identity. At the post-project evaluation phase, project narratives about value created for society, realised project outputs and outcomes, awards and achievements are crafted and become promoted widely to a public. This creates a sense of pride of project work and achievements that we will discuss in Chapter 7.

There is a continuous process of sustaining and repeating the appropriate project narratives in the dynamic flux of project activities. Project leaders tend to strive for consistency in communicating project narratives. However, project narratives are temporal in nature meaning that they tend to change throughout the project lifecycle. Project narratives get (re)iterated to many different audiences (internal and external stakeholder) and re-stated in many different ways throughout a project lifecycle.

2.7 Conclusion

In this chapter, we discuss the importance of narratives in megaprojects. We start by describing what megaprojects are and why the number and size of megaprojects are on the rise due to the technological, political, economic, aesthetic and community pride sublime. Making an identity is critical for megaprojects as even though the number and size of these projects increase worldwide, these projects are shrouded in inefficiencies and failures. A powerful megaproject identity can stimulate the diverse stakeholders to commit themselves to the megaproject. We highlight how identity and sensemaking are related and why narratives are required for sensemaking in megaprojects. We discuss a project image narrative and how it is used and promoted in megaprojects. We then describe how counter-narratives are also present in megaprojects along with narratives and the role of dominant narrative in creating an identity which is essential for the success of megaprojects. We also show that different project narratives can be identified depending on the specific project nature. We present project narratives that maybe observed throughout a project lifecycle. In the next chapter, narratives and counter-narratives by promoters and protesters in megaprojects will be discussed.

CASE STUDY 2: THAMES TIDEWAY, UNITED KINGDOM

Tideway is a major part of the overall upgrade of the London's Victorian sewerage systems. The project involves building a 25-km Thames Tideway Tunnel under London's River Thames to prevent the discharge of the tens of millions of tonnes of untreated waste that currently pollute the river. The scheme is the biggest investment in London's sewage system since 1875 and the largest water infrastructure megaproject in the UK. Thames Water, London's water utility, created a new independent regulated company, Bazalgette Tunnel Ltd., to be the owner and operator of Tideway, financed largely by pension funds because the project was too big for its own balance sheet. Construction started in 2016 and it is expected to be completed by 2025. There are 24 construction sites in London, spanning from Acton in the west London to Beckton in the east. Tideway is Bazalgette's brand for the megaproject; it does not include the word "tunnel" in order to highlight the project mission and its outcome, which is to address the problem of discharging waste matter into the tidal River Thames, rather than the physical output.

The image-shaping narrative of Tideway was always strong as an environmental improvement project: 1) to protect the Thames Tideway ecosystem; 2) to reduce pollution from sewage-derived litter; and 3) to protect the health of recreational river users. However, the delivery identity narrative was missing. Tideway therefore launched the "Our Space" initiative, which asked employees to reflect on what they were doing on this megaproject in that open meeting space. This included a whiteboard wall map of the River Thames on which to brainstorm ideas, which generated many keywords and a strong theme emerged as "reconnecting Londoners with the River Thames". This identity narrative was captured in a cartoon of all the activities enabled for Londoners on the Thames.

During the early months of project delivery, the CEO of Tideway spent 50 per cent of his time on this crafting, which played a crucial role in forming the megaproject's identity narrative. He has also delivered a number of presentations about Tideway and its identity at various external industry and academic events. The CEO is perceived as an inspirational leader who is reflective and a good storyteller. He is outlined his approach to the job as follows:

> I was keen to champion early career professionals to form a collective to consider what they might like to do to further their progress, but also to contribute to the things that we do in the industry. And almost to encourage them to challenge and to question and to really allow them to believe it is OK as a less experienced person to say: "Why do we do this?" My worries are by the time you get into a position of authority when you can set directions and you can make changes you have forgotten half of the things; you now become blind to some of the things that always were logical. You lose some of that natural inquisitiveness that you have when you are younger. Things that do not make sense and you question it. I was keen to encourage younger people to be confident enough, and to work as a collective to find ways of contributing to changes and how we do things.

The CEO was also keen to encourage early career professionals to be heavily engaged with the innovation programme at Tideway. Tideway is the first megaproject to use the UK industry-wide i3P (Infrastructure Industry Innovation Partnership) online innovation portal. The i3P portal builds on the success of Innovate 18, Crossrail's innovation programme which contains over 400 innovations over a three-year period. i3P's vision is to establish a driver for innovation in the UK infrastructure sector and to provide a mechanism for strategically directing innovation to address the challenges facing the sector. I3P's vision is to create a "safe space" to identify areas for industry improvement, share innovative ideas and enable members to collaborate on projects that drive value, and also to establish a collaborative culture of innovation across the owner and supplier domains. Tideway has established a team of innovation managers to manage its innovation programme.

The approach to innovation in Tideway is to encourage and stimulate innovation across the supplier domain. There are multiple reasons for doing so. Firstly, the research for novel ideas aims to increase project performance in terms of schedule and budget savings. Secondly, it aims to create beneficial effects for stakeholder communities, making construction works safer, shorter, and less disruptive. Thirdly, innovation is seen as a catalyst to create a world-class workforce which benefits not only Tideway, but also the overall sector.

Finally, by proving capability to deliver projects more quickly and effectively through innovation, Tideway aims to attract future investments.

Tideway was granted access to Crossrail's Innovate 18 portal, where it gathers ideas and learns from Crossrail's experiences. A number of best practices were captured from Crossrail, such as the central budget model to fund ideas and the network of innovation champions. Tideway developed a slightly different governance from Crossrail, consisting of an 'organisation within the organisation' that benefits from a stand-alone budget dedicated to supporting innovation. Not only innovations that can deliver direct benefits to Tideway are selected, but out-of-scope innovations are also considered and shared if deemed valuable to other organisations within Tideway's network. Compared to Crossrail, another innovative element of Tideway's strategy is the innovation network where it is developing relationships with organisations outside the construction sector so that Tideway can benefit from the application of existing practices or technologies from other sectors.

In order to align participants' interests, support integration, and motivate collaborative behaviours, alliance agreements were developed to enable loss and gain sharing across the commercial interface. These support the innovation journey, and apply to the Optimised Contractor Involvement process, where nominated suppliers are asked to provide inputs to improve the current design with the aim of driving efficiency and innovation into the design and also construction methods. Different forms of incentives and reward schemes are used to motivate innovative ideas from both the suppliers and the owner organisation itself. Yet, in addition to formal incentives, leadership was needed to create the right culture to support the innovation process, helping people developing a natural inclination to creative thinking and, simultaneously, ensuring tolerance of failure. Leading by example and storytelling were deemed to be instrumental to achieve these objectives.

To celebrate the spirit of Tideway innovation, in June 2017 it hosted the Tideway Innovation Forum at the original studios where the BBC TV series *Dragons' Den* is filmed as they sought to "Imagine, Innovate and Inspire" a reconnection with the River Thames. They welcomed new ideas seeking £10,000 and £20,000 of investment which were likely to improve Tideway delivery, or secure a legacy of innovation, from across their Alliance and stakeholders. The Tideway innovation team assisted innovators to develop their ideas and prepare a business case. With £100,000 available on the day, the best ideas were pitched to the Tideway Dragons: four senior leadership team members. From 30 innovative ideas submitted to the Great Think – an innovation community to discuss, discover and share ideas hosted by i3P – in the run-up to the event, the innovation team and champions across the Alliance whittled entries down to the five ideas given the opportunity to pitch to the Den. Each innovation champion had three minutes to pitch their idea, and the Dragons just ten minutes to interrogate them before making a decision. One example of an unsuccessful idea pitched was to develop Augmented Reality (AR) technology for remote surveys and inspections. Whilst this idea was unsuccessful, the Dragons recognised the future potential and so this idea was passed to i3P for further discussion at a future Innovation Forum. An example of a successful idea was to develop and trial the 'Rightway Bootcamp', an industry-first 11-week induction programme covering physical, mental and financial health supporting Tideway's aim to have everyone leave the project healthier than when they arrived.

Tideway also held engagement campaigns within and outside the company. Colleagues across the Alliance continued to develop novel solutions, implement innovative technologies, and share their knowledge on the i3P portal, with 50 innovations published online, and 80 new ideas submitted to innovation team. Tideway continues to engage across the industry to support collaboration and innovation. This includes the CEO chairing i3P and hosting its first live funding event, i3P Spark, which had over 30 collaborative submissions from member organisations. £100,000 from i3P Collaborative Innovation Fund was invested in the three winning pitches from the final in ideas deemed to have the highest potential to benefit the whole sector.

Exercise

From the above case study please answer the following questions:

1) Why is a project internal identity narrative needed, and how one was created in relation to the Tideway project?
2) To what extent is the innovation programme in Tideway robust, and what else could be done as part of the programme?
3) How and why is the project external image narrative important throughout the project lifecycle?

Sources

Interviews with CEO and innovation team and desktop research.
Stride, P. (2019). *The Thames Tideway Tunnel: Preventing another great stink*. Stroud, UK: The History Press.
The case is published in Winch, G. M., Maytorena-Sanchez, E., & Sergeeva, N. (2022). *Strategic project organizing*. Oxford University Press.

References

Albert, S., & Whetten, D. A. (1985). Organizational identity. In L. L. Cummings & B. M. Staw (eds.), *Research in organizational behavior* (pp. 263–295). Greenwich, CT: JAI Press.

Andrews, M. (2002). Introduction: Counter-narratives and the power to oppose. *Narrative Inquiry*, 12(1), 1–6.

Andrews, M. (2004). Opening to the original contributions: Counter-narratives and the power to oppose. In M. G. W. Bamberg, & M. Andrews (eds.), *Considering counter-narratives: Narrating, resisting, making sense* (pp. 1–6). Philadelphia, PA: John Benjamins.

Ancona, D. (2011). Sensemaking. Framing and acting in the unknown. In: N. Nohria, S. Snook, & R. Khuranales (eds.), *The handbook for teaching leadership. Knowing, doing and being* (pp. 3–21). Thousands Oaks, CA: Sage.

Ashforth, B. E., Harrison, S. H., & Corley, K. G. (2008). Identification in organizations: An examination of four fundamental questions. *Journal of Management*, 34(3), 325–374.

Ashforth, B. E., & Humphrey, R. H. (1997). The ubiquity and potency of labeling in organizations. *Organization Science*, 8(1), 43–58.

Boje, D. M. (2001). *Narrative methods for organizational and communication research*. London: Sage.

Boddy, D., & Paton, R. (2004). Responding to competing narratives: Lessons for project managers. *International Journal of Project Management*, 22(3), 225–233.

Boyce, M. E. (1995). Collective centring and collective sense-making in the stories and storytelling of one organization. *Organization Studies*, 16(1), 107–137.

Bresnen, M., & Marshall, N. (2001). Understanding the diffusion and application of new management ideas in construction. *Engineering, Construction and Architectural Management*, 8(6/5), 335–345.

Bucholtz, M., & Hall, K. (2005). Identity and interaction: A sociocultural linguistic approach. *Discourse Studies*, 7(4-5), 585–614.

Christensen, L. T. (1995). Buffering organizational identity in the marketing culture. *Organization Studies*, 16(4), 651–672.

Cicmil, S., Williams, T., Thomas, J., & Hodgson, D. (2006). Rethinking project management: Researching the actuality of projects. *International Journal of Project Management*, 24(8), 675–686.

Corley, K. G., & Gioia, D. A. (2004). Identity ambiguity and change in the wake of a corporate spin-off. *Administrative Science Quarterly*, 49(2), 173–208.

Davies, A., Gann, D., & Douglas, T. (2009). Innovation in megaprojects: Systems integration at London Heathrow terminal 5. *California Management Review*, 51(2), 101–125.

Davies, A., & Mackenzie, I. (2014). Project complexity and systems integration: Constructing the London 2012 Olympics and Paralympics games. *International Journal of Project Management*, 32(5), 773–790.

Delgado, R. (1995). Legal storytelling: Storytelling for oppositionists and others: A Plea for narrative. In R. Delgado (ed.), *Critical race theory: The cutting edge* (pp. 64–74). Philadelphia, PA: Temple University Press.

Denis, J. L., Langley, A., & Rouleau, L., (2007). Strategizing in pluralistic contexts: Rethinking theoretical frames. *Human Relations*, 60(1), 179–215.

Drevin, L., & Dalcher, D. (2011). Antenarrative and narrative: The experiences of actors Involved in the development and use of information systems. In D. M. Boje (ed.), *Storytelling and the future of organizations: An antenarrative handbook* (pp. 148–162). New York: Routledge.

Escalas, J. E. (2007). Self-referencing and persuasion: Narrative transportation versus analytical elaboration. *Journal of Consumer Research*, 33(4), 421–429.

Esposito, G., Nelson, T., Ferlie, E., & Crutzen, N. (2021). The institutional shaping of global megaprojects: The case of the Lyon-Turin high-speed railway. *International Journal of Project Management*, 39(6), 658–671.

Esposito, G., & Terlizzi, A. (2019). The clash of worlds in global megaprojects: Policy narratives and counter-narratives in the case of the Lyon-Turin high-speed railway. In Società Italiana di Scienza Politica Conference, University of Salento, Lecce, Italy.

Flyvbjerg, B. (2003). The lying game. *EuroBusiness*, 5(1), 60–62.

Flyvbjerg, B. (2007). Policy and planning for large-infrastructure projects: Problems, causes, cures. *Environment and Planning B: Planning and Design*, 34(4), 578–597.

Flyvbjerg, B. (2014). What you should know about megaprojects and why: An overview. *Project Management Journal*, 45(2), 6–19.

Flyvbjerg, B. (2017). *The Oxford handbook of megaproject management*. Oxford, UK: Oxford University Press.

Frandsen, S., Kuhn, T., & Lundholt, W. (2017). *Counter-narratives and organization*. New York and London: Routledge.

Gil, N. A. (2010). Language as a resource in project management: A case study and a conceptual framework. *IEEE Transactions on Engineering Management*, 57(3), 450–462.

Gil, N. A., & Pinto, J.K. (2018). Polycentric organizing and performance: A contingency model and evidence from megaproject planning in the UK. *Research Policy*, 47(4), 717–734.

Gioia, D. A., & Chittipeddi, K. (1991). Sensemaking and sensegiving in strategic change initiation. *Strategic Management Journal*, 12(6), 433–448.

Gioia, D. A., Patvardhan, S. D., Hamilton, A. L., & Corley, K. G. (2013). Organizational identity formation and change. *The Academy of Management Annals*, 7(1), 123–193.

Gioia, D. A., Schultz, M., & Corley, K. G. (2000). Organizational identity, image and adaptive instability. *Academy of Management Review*, 25(1), 63–81.

Grayson, K. (1997). Special session summary narrative theory and consumer research: Theoretical and methodological perspectives. In M. Brucks & D. J. MacInnis (eds.), *NA – Advances in consumer research* (Vol. 24, pp. 67–70). Provo, UT: Association for Consumer Research.

Harper, S. R. (2009). Niggers no more: A critical race counter-narrative on Black male student achievement at predominantly White colleges and universities. *International Journal of Qualitative Studies in Education*, 22(6), 697–712.

Harris, A., Carney, S., & Fine, M. (2001). Counter work: Introduction to 'Under the covers: Theorising the politics of counter stories'. *International Journal of Critical Psychology*, 4, 6–18.

Havermans, L. A., Keegan, A., & Den Hartog, D. N. (2015). Choosing your words carefully: Leaders' narratives of complex emergent problem resolution. *International Journal of Project Management*, 33(5), 973–984.

Hietajärvi, A. M., & Aaltonen, K. (2018). The formation of a collaborative project identity in an infrastructure alliance project. *Construction Management and Economics*, 36(1), 1–21.

Humphreys, M., & Brown, A. D. (2002). Narratives of organizational identity and identification: A case study of hegemony and resistance. *Organization Studies*, 23(3), 421–447.

Kodeih, F., & Greenwood, R. (2014). Responding to institutional complexity: The role of identity. *Organization Studies*, 35(1), 7–39.

Lindgren, M., & Packendorff, J. (2007). Performing arts and the art of performing – On co-construction of project work and professional identities in theatres. *International Journal of Project Management*, 25(4), 354–364.

Maitlis, S., & Christianson, M. (2014). Sensemaking in organizations: Taking stock and moving forward. *Academy of Management Annals*, 8(1), 57–125.

Maitlis, S., & Lawrence, T. B. (2007). Triggers and enablers of sensegiving in organizations. *Academy of Management Journal*, 50(1), 57–84.

Mathur, S., Ninan, J., Vuorinen, L., Ke, Y., & Sankaran, S. (2021). An exploratory study of the use of social media to assess benefits realization in transport infrastructure projects. *Project Leadership and Society*, 2, 1–10.

McLean, K. C. (2015). *The co-authored self: Family stories and the construction of personal identity*. New York, NY: Oxford University Press.

McLean, K. C., & Syed, M. (2015). Personal, master, and alternative narratives: An integrative framework for understanding identity development in context. *Human Development*, 58(6), 318–349.

McQuillan, M. (2000). *The narrative reader*. London and New York: Routledge.

Morris, P. W., & Pinto, J. K. (2010). *The Wiley guide to project organization and project management competencies*. New Jersey: John Wiley & Sons.

Ninan, J., Mahalingam, A., & Clegg, S. (2020). Power and strategies in the external stakeholder management of megaprojects: A circuitry framework. *Engineering Project Organization Journal*, 9(1), 1–20.

Ninan, J., & Sergeeva, N. (2021). Labyrinth of labels: Narrative constructions of promoters and protesters in megaprojects. *International Journal of Project Management*, 39(5), 496–506.

Ninan, J., Mahalingam, A., & Clegg, S., (2022). Power in news media: Framing strategies and effects in infrastructure projects. *International Journal of Project Management*, 40(1), 28–39.

Olander, S., & Landin, A. (2008). A comparative study of factors affecting the external stakeholder management process. *Construction Management and Economics*, 26(6), 553–561.

Packendorff, J. (1995). Inquiring into the temporary organization: New directions for project management research. *Scandinavian Journal of Management*, 11(4), 319–333.

Phillips, N., & Oswick, C. (2012). Organizational discourse: Domains, debates, and directions. *Academy of Management Annals*, 6(1), 435–481.

Pontikes, E. G. (2012). Two sides of the same coin: How ambiguous classification affects multiple audiences' evaluations. *Administrative Science Quarterly*, 57(1), 81–118.

Pontikes, E. G. (2018). Category strategy for firm advantage. *Strategy Science*, 3(4), 620–631.

Priemus, H., Flyvbjerg, B., & van Wee, B. (2008). *Decision-making on mega-projects: Cost-benefit analysis, planning and innovation*. Cheltenham: Edward Elgar Publishing.

Reed, A. H., & Knight, L. V. (2010). Effect of a virtual project team environment on communication-related project risk. *International Journal of Project Management*, 28(5), 422–427.

Roehrich, J., & Lewis, M. (2014). Procuring complex performance: Implications for exchange governance complexity. *International Journal of Operations & Production Management*, 34(2), 221–241.

Schultz, M., & Hernes, T. (2013). A temporal perspective on organizational identity. *Organization Science*, 24(1), 1–21.

Sergeeva, N. (2017). Labeling projects as innovative: A social identity theory. *Project Management Journal*, 48(1), 51–64.

Sergeeva, N. (2019). *Making sense of innovation in the built environment*. Oxfordshire, UK: Routledge.

Sergeeva, N., & Ninan, J. (2022). Project narratives: Directions for research. In G. M. Winch, M. Brunet, & D. Cao (eds.), (in press). https://www.e-elgar.com/shop/gbp/research-handbook-on-complex-project-organizing-9781800880276.html

Sergeeva, N., & Roehrich, J. (2018). Learning in temporary multi-organizations: Constructing identities to realize performance improvements. *Industrial Marketing Management*, 75, 184–192.

Sergeeva, N., & Winch, G.M. (2021). Project narratives that potentially perform and change the future. *Project Management Journal*, 52(3), 264–277.

Smircich, L., & Morgan, G. (1982). Leadership: The management of meaning. *The Journal of Applied Behavioral Science*, 18(3), 257–273.

Sorsa, V., & Vaara, E. (2020). How can pluralistic organizations proceed with strategic change? A processual account of rhetorical contestation, convergence, and partial agreement in a nordic city organization. *Organization Science*, 31(4), 839–864.

Söderlund, J., Sankaran, S., & Biesenthal, C. (2017). The past and present of megaprojects. *Project Management Journal*, 48(6), 5–16.

Squire, C. (2002). White trash pride and the exemplary black citizen: Counter-narratives of gender, "race" and the trailer park in contemporary daytime television talk shows. *Narrative Inquiry*, 12(1), 155–172.

Sturup, S. (2009). Mega projects and governmentality. *World Academy of Science, Engineering and Technology*, 3(6), 892–901.

Todorov, P. (1971). *The poetics of prose*. Oxford: Blackwell.

Vaara, E., & Tienari, J. (2011). On the narrative construction of multinational corporations: An antenarrative analysis of legitimation and resistance in a cross-border merger. *Organization Science*, 22(2), 370–390.

Van Marrewijk, A. (2007). Managing project culture: The case of Environ Megaproject. *International Journal of Project Management*, 25(3), 290–299.

Veenswijk, M., & Berendse, M. (2008). Constructing new working practices through project narratives. *International Journal of Project Organisation and Management*, 1(1), 65–85.

Veenswijk, M., Van Marrewijk, A., & Boersma, K. (2010). Developing new knowledge in collaborative relationships in megaproject alliances: Organising reflection in the Dutch construction sector. *International Journal of Knowledge Management Studies*, 4(2), 216–232.

Volker, L. (2019). Looking out to look in: inspiration from social sciences for construction management research. *Construction Management and Economics*, 37(1), 13–23.

Watson, T. J., & Bargiela-Chiappini, F. (1998). Managerial sensemaking and occupational identities in Britain and Italy: The role of management magazines in the process of discursive construction. *Journal of Management Studies*, 35(3), 285–301.

Weick, K. E. (1995). *Sensemaking in organizations*. Sage. London.

Weick, K. E., Sutcliffe, K. M., & Obstfeld, D. (2005). Organizing and the process of sensemaking. *Organization Science*, 16(4), 409–421.

Winch, G. M., Maytorena-Sanchez, E., & Sergeeva, N. (2022). *Strategic project organizing*. London: Oxford University Press.

Zilber, T. B. (2007). Stories and the discursive dynamics of institutional entrepreneurship: The case of Israeli high-tech after the bubble. *Organization Studies*, 28(7), 1035–1054.

3

PROMOTERS AND PROTESTERS' NARRATIVES

3.1 Introduction

Megaprojects involve multiple stakeholders in the form of sponsors, partners, experts, contractors, government agencies, opposing stakeholders and other institutions of external players, and are rarely built with in-house resources (Flyvbjerg, 2014; Miller et al., 2017). Broadly, these stakeholders can be classified as promoters and protesters. In this chapter we discuss the different stakeholders, their classification into promoters and protesters, and the differences in the narratives they create.

3.2 Stakeholders in megaprojects

Both internal and external stakeholders are important actors in megaprojects who affect their outputs and outcomes (Datta et al., 2020). Mitchell et al. (1997) define stakeholders broadly as anyone that can have an impact on an organisation's actions or who experiences an impact as a result of them. Stakeholders are in a position to influence the wellbeing of an organisation, defined in terms of its capacity to achieve goals (Freeman, 1984). Since many organisations are responsible for the success of projects, stakeholder management is significant in the project context (Achterkamp & Vos, 2008). Stakeholder management in project management scholarship involves bringing stakeholder concerns to the surface and developing robust stakeholder relationships in complex project environments (Bourne & Walker, 2005). In their review of stakeholder literature in projects, Littau et al. (2010) note that stakeholders can be classified in three ways: (1) those who have an interest in the project; (2) those who can affect the project; and (3) those that both have an interest in and can affect the project. In contrast to smaller projects, megaprojects, due to their larger scale, have a greater number of stakeholders.

DOI: 10.1201/9781003248378-3

Complexity in megaprojects stem from the institutional differences, such as divergent perceptions regarding the legitimate means and ends of the project rather than the mere number of different organisational entities or stakeholders involved in the megaproject (Orr & Scott, 2008). In their study of infrastructure projects, Li et al. (2005) underline that the interest of each stakeholder is different. For example, the interests for road alignment for the project team would be minimising the cost and time of construction, whereas local vendors near the project would be in safeguarding their land, and users would be safer and comfortable travel. In addition, different utility agencies such as water and electricity will have different interests in the highway project, based on their assets in the area. Li et al. (2012) investigated the different priorities of stakeholders in an infrastructure project in Hong Kong. They found out that while government drafts the potential economic benefits for the project, the community is focused on sustainable land use, the project-affected groups are focused on tangible compensation and the pressure groups are concerned with the ecological impacts of the project. Thus, the construction industry in general confronts many more conflicts than most other industries, in part due to the structurally conflicting interests of various project parties over matters as fundamental as cost, quality and schedule (Black et al., 2000) and the lack of a common rationale and culture binding all project participants and stakeholders (Vrijhoef & Koskela, 2000).

The decision-making also becomes complex and fuzzy with many stakeholders and their diverse vested interests. A study conducted by Bekdik and Thuesen (2016) on the decisions surrounding 27 hospital infrastructure projects in Denmark revealed that the patient room, which has certain guidelines, are differently designed in each case. Their analysis of the stakeholders showed that each hospital had a different set of active stakeholders, and hence the priorities and decisions varied from case to case. Decisions regarding the rooms were taken by different stakeholders, according to who was active in each hospital. Thus, the results of a project are dependent on the interests of the stakeholders active in that project context.

Stakeholders can include customers, suppliers, employees, financiers and communities (Dunham et al., 2006). While internal stakeholders, such as the project builders, have a contractual relationship with the client, external stakeholders have no such relationship and rely on regulators, political influence or public campaigns to enforce a claim (Winch, 2007). External stakeholders include the stakeholders peripheral to the project, such as the owners of the land on which the project is to be built, those who are inconvenienced by the construction noises, vibrations, diversions, etc., and those who stand to benefit from the project improving the services (Ninan et al., 2020; Viitanen et al., 2010). For most of the external stakeholders, megaprojects are disruptions to their everyday life. Ignoring the needs and expectations of the external stakeholders can generate social unrest or community resistance through collective action against the project (Liu et al., 2018; van den Ende & Van Marrewijk, 2019) in the form of petitions, boycott, strikes, protests, picketing or even vandalism (Oppong et al., 2017). Mok et al. (2015) note that conflicts or resistance from the public can adversely affect or even kill the project despite the public being an external stakeholder who lacks a formal project authority. These external stakeholders seek to shape major megaproject decisions, including budget and scope, in accord with specific vested interests and are also referred to as the 'stakeholders of the shadows' (Winch, 2017). Thus, as Smith and Love (2004) record, successful management of external stakeholders can result in the reduction in waste of effort, time and resources in project management.

These projects involve a multiplicity of public and private, internal and external stakeholders with different interests, values and rationality (Van Marrewijk, 2015). The conflicting interests can be broadly classified into either support for or resistance against the megaproject. Hence, they can be classified broadly as either promoters or protesters of the project.

3.2.1 Promoters

The promoters or proponents include multiple stakeholders who share an interest in supporting the completion of a megaproject. For example, promoters advocate for funding large-scale projects to create long-term economic benefits. They can be different stakeholders, such as government agencies, contractors and lobby groups actively campaigning for the megaproject. They invest in megaprojects in order to stimulate the economy or in support of their other interests. Ninan and Sergeeva (2021), in their study of the High Speed 2 (HS2) project in the UK, note a number of different promoter groups: the UK government, the project spokesperson, the Association of Train Operating Companies (ATOC), etc. The promoters aim to create a narrative in support of the project. Megaprojects and those who lead them, internal stakeholders, are in favour of promoters and their supportive narratives.

3.2.2 Protesters

The protesters of a megaproject also include a range of stakeholders who are interested in derailing the project. These include stakeholders, such as people affected by the project, activists and non-governmental organisations. Other protesters, like local communities, may be against the project for various reasons, including high cost, environmental effects and the impact on people. Different protester groups, such as opposition party MPs, members of interest groups, action groups, etc., were seen in the case of the HS2 project in the UK (Ninan & Sergeeva, 2021). The protesters aim to create a narrative against the project. Megaprojects (and those who lead them) work on ways and strategies of overcoming protesters' narratives and possibly turning protesters and their narratives into supporters who accept their promotional narratives. In cases when protesters' narratives become dominant and overcome promoters' narratives a project is more likely to be cancelled.

3.2.3 Conflict between promoters and protesters

There is an ongoing process of dynamic interaction between the promoter and the protester groups of stakeholders (Ninan & Sergeeva, 2021) as the sources of difference, potential and actual conflict are exacerbated in megaprojects (Flyvbjerg, 2014) because of their increased scale, duration, complexity and a wider range of external stakeholders. Miller et al. (2017) note that in megaprojects, when decisions are to be made on the specifics of the services, technical solution and financials, projects are subject to conflicting pressures and risks and turbulence emerge. Conventional project control systems, well suited to more routine and standardised projects, cannot guarantee success in conditions of chronic uncertainty, complexity and contingency, as has been noted by authors such as Flyvbjerg et al. (2003). Megaprojects combine multiple competing partners with different interests, values and rationality (Van Marrewijk, 2015). This creates pluralistic sites in which power is diffused among distributed actors and conflicting institutional logics are evident (Biesenthal et al., 2018).

Since these projects take a long period of time to come to fruition, stakeholders may adopt a range of differing strategies over the course of its lifespan to propel their vested interests (Aaltonen & Sivonen, 2009; Flyvbjerg, 1998; Ninan et al., 2020). In such situations of high ambiguity, megaprojects strive to create an organisational identity (Sergeeva & Roehrich, 2018), as discussed in Chapter 2. Constructing a favourable identity is of paramount importance to bring together the promoters and protesters, as the inability to garner the legitimacy and support of external stakeholders can affect the delivery of the infrastructure asset (Ninan et al., 2020). Narratives can arise from different actors and no one narrative can be considered as true as there are "as many narratives as there are actors" (Cooren, 1999). In an ongoing flow of plausible and supportive and oppositional, counter-narratives by promoters and protesters, megaproject executives and managers work on crafting and maintaining a dominant, consistent and coherent narrative around a project that mobilises resources in order to achieve the desired outcomes. They feel committed to crafting and maintaining a consistent narrative about a project throughout its lifecycle.

3.2.4 Resolution of conflicts

Stakeholder theory, as outlined by Donaldson and Preston (1995), has four parts: descriptive, which describes what organisations actually do; instrumental, which focuses on the outcomes of managerial behaviours; normative, which provides guidance on what organisations should do; and managerial, which speaks to the needs of the customers. Of these, the most commonly used are the normative and instrumental views on stakeholder theory. Henisz et al. (2014) record that in the normative view there is only a moral management with no real returns, while in the instrumental view there is a focus on company image with increased returns. Noland and Phillips (2010) explain instrumental and normative view as, respectively, strategic and moral, based on the goal, manner and method of management. Investing in the company image can alter stakeholder behaviour, generate shareholder value and ensure that the business plan will proceed both on schedule and on budget (Freeman, 1984). Instrumental stakeholder theory focuses on managing stakeholders for achieving an organisation's corporate objectives (Donaldson & Preston, 1995). Thus, stakeholder engagement stems from the normative perspective and stakeholder management stems from the instrumental perspective.

Strategies such as adaptation, compromise, negotiation, concession and avoidance are used for managing stakeholders in project settings (Chinyio & Akintoye, 2008). However, Di Maddaloni and Davis (2017) note that despite attempts by projects to adopt these strategies, stakeholders are often adversarial and hence most projects lack the 'reservoir of support' from the community. It should also be noted that the most affected stakeholder may not necessarily be the most vocal (Van Marrewijk et al., 2008) and the activities of the vocal few can result in the project not delivering on its intended benefits. In contrast to the normative view, strategies such as marketing (Turner et al., 2019) and branding (Ninan et al., 2019) can help projects focus on the project image and ensure that the project will proceed as planned. In project settings, Derakhshan et al. (2019) explain instrumental stakeholder theory as managing stakeholders for their role in maximising organisation's benefits, rather than because of their legitimate rights. One way of managing stakeholders for maximising organisational benefits is through discourses.

There are different types of discourses for managing stakeholders. Discourses for moral management includes communication solely for the sake of reaching agreement rather than in order to pursue any particular interests (Noland & Phillips, 2010). Such communication should be uncorrupted by power differences and strategic motivations. In contrast, discourses for strategic management have a strategic intent with a focus on achieving an organisation's corporate objectives (Zakhem, 2007). They are undertaken with strategic, though not necessarily intentionally dishonest or malicious, motivations (Noland & Phillips, 2010). In project settings, Ninan et al. (2020), using organisational power theories, differentiate these two types of management in an infrastructure megaproject's use of social media. Social media for persuasion involved the project reaching out to the community for mutual agreements and can be categorised as moral management. In contrast, social media for framing and hegemonising involved the project using strategic discourses to influence stakeholders to move towards the project's interests and can be categorised as strategic management. Strategic discourses aimed at external stakeholders can also affect the project team rationalities and decision-making as these discourses percolate and 'trickle down' to the internal stakeholders (Ninan et al., 2022). Here, we are interested in the strategic management of external stakeholders using narratives.

Stakeholder theory is often highlighted as a 'genre' as it includes a number of theories and a range of applications, all of which have stakeholders at the centre. The theory focuses on a broad array of disciplines, including business ethics, corporate strategy, finance, accounting, management and marketing (Parmar et al., 2010).

Within these, a focus on marketing involves developing marketing theory and practice along stakeholder theory lines (Roper & Davies, 2007), which has implications for the instrumental view of stakeholders. In the case of projects, it is important that the core narrative should be stable and promoted because that is the reason the project is there. A stable narrative is necessary to ensure that the objectives of the project do not change during the pre-construction or construction phases (Sergeeva & Winch, 2021). Narratives are defined as conversations, dialogues and stories that communicate a phenomenon (Garud & Turunen, 2017). Narratives are cultural mechanisms that refer to a set of events and the contextual details surrounding their occurrence (Bartel & Garud, 2009). People make sense of their lives via narrative thought as the temporal and dramatic dimension of human existence is emphasised in them (Polkinghorne, 1991). They organise their experiences and create order out of random incidents and events (Grayson, 1997). Such order helps people describe and understand the passage of events (Ricoeur, 1991). Within projects, narratives determine how the project team deals with emergent problems and even how projects are perceived by others (Enninga & van der Lugt, 2016; Havermans et al., 2015).

Ganz (2011) notes that what required to enable progress in situations of uncertainty and conflicts is command over the 3 H's – the Hands, the Head and the Heart. The Hands represent action, of either learning, adapting, or mastering novel skills. The Head represents strategy, which involves imagining how to transform one's resources to enable an action. Finally, the Heart represents motivation either of the urgency of the need to act, or the hope for success, and the courage to go after the action. Thus, collective action is not just hands or actions, such as the use of participatory stakeholder engagement, citizen report cards, public expenditure tracking, or social audits, but also the values and identity at play shaped by the heart. Similar to this, Aristotle differentiates between logos and ethos. While logos is the logic of the argument, pathos is the feeling the argument evokes. Both logos and ethos are required for the resolution of conflicts and collective action. Language and narratives can enable a transformation of the heart to enable the head and give purpose to the hands.

Language is at the very centre of project organising as it helps construct project events instead of just representing them (Havermans et al., 2015). Language, communication and storytelling more specifically all play significant roles in increasing legitimacy and acquiring resources (Lounsbury & Glynn, 2001). Therefore, projects can be understood as social constructions that are produced and re-produced in its everyday narrative interactions (Lindgren & Packendorff, 2007). Project narratives are important vehicles through which meanings are negotiated, shared and contested (Veenswijk & Berendse, 2008). They help in creating a shared memory or a collective brand image at the levels of firm and sector as a whole (Duman et al., 2018). For example, the Environ megaproject employees considered in the work by Van Marrewijk (2007) were able to strongly identify themselves as innovative and entrepreneurial when they were referred to as 'Gideon's gang'. This is a biblical metaphor for a brave group of fearless men who use creative, innovative methods to reach their goals. Explaining this further, Havermans et al. (2015) note that whether the project is described as 'routine' or 'ground-breaking' or whether those with opinions on the project are described as 'nuisance' or 'an important source of new ideas' are dependent on how they are framed by leaders. While coherent and consistent project narratives are required for the survival of the project, they are contested by different agencies across the life cycle of the project (Boddy & Paton, 2004). It is here that we situate this research to understand the dynamics through which megaproject narratives are mobilised in the external stakeholder environment.

Polkinghorne (1991) notes that how people organise is dependent on the cues emanating from external perpetual senses, internal bodily sensations, and cognitive memories. He argues that narratives are one of the main cognitive organising processes as they give meaning to temporal events by identifying them as parts of a plot. Similarly, Rappaport (2000) claims that narratives can be treated as a 'cognitive instrument' as they can impact subject's thinking and emotional life. Narratives can be targeted at audiences and can have both performative and strategic implications, thereby making them effective in constructing organisational identities (Dailey & Browning, 2014; Sergeeva, 2019; Sergeeva & Winch, 2021). Thus, it is possible for narratives to mobilise and bring people together (Duman et al., 2018). From a governance perspective, Abolafia (2010) records how policy makers employ plotted, plausible and repeated narratives to shape the reaction of people to the changes occurring around them. Narratives are powerful mechanisms for translating ideas across the organisation so that they are comprehensible and appear legitimate to others (Bartel & Garud, 2009). Narrative is the discursive means that equip us to access values and provides the courage to make choices under conditions of uncertainty and to exercise agency (Ganz, 2011). Vaara et al. (2016) claim that narratives are mobilised in many ways as part of discourses and communication. In the next section, we discuss the different narratives promoters and protesters mobilise in megaprojects.

3.3 Promoters and protesters narratives

There are as many narratives as there are actors (Cooren, 1999). Since there are broadly two sets of stakeholders – promoters and protesters – we discuss the promoters' and protesters' narratives in different aspects of megaprojects, such as the need for the project, the stakeholder consultation process, environmental sustainability and noise.

3.3.1 Need for the project

The need for the megaproject is one of the most critical narratives resulting in whether or not the project is built. While the promoters of the megaproject aim to create a narrative that the project is needed, the protesters of the megaproject strive to propel a narrative that the project is unnecessary.

Stressing the economic, social or environmental benefits that the megaprojects would achieve for society is one of the ways to create a favourable narrative. Protesters against the megaproject advance a counter-narrative by highlighting that the megaproject does not fare well in terms of supplying these benefits. They attack the economic feasibility of the project and claim that the business case of the project does not stack up. Such counter-narratives destabilise the dominant narrative, i.e., that the project is needed, to a new narrative that the project is not needed. The promoters of the project destabilise the opposition narrative, that the project is not needed, by highlighting that, irrespective of the counter-narrative, the project is still needed. There are different strategies to advance a narrative or destabilise a narrative, which this book will consider in the subsequent chapters. Thus, there are multiple, often conflicting narratives of the need for the megaproject, and they develop in a longitudinal way countering the existing dominant narrative.

3.3.2 The stakeholder consultation process

Narratives can even be initiated by the protesters of the project, and a narrative of the stakeholder consultation process is a prime example of this. The protesters often campaign that an effective stakeholder consultation process was not done, and that the government is not hearing the community's concern regarding the project.

The protesters of the project emphasise that the consultation process was not a fair and open debate about the project. To counter this narrative, the promoters give details of the consultation process and call it one of the largest consultations ever undertaken by a government. They even claim that the project had a fair debate with tens of thousands of people and that the consultation process was carried out properly. By emphasising that the project has conducted multiple consultation events throughout the project, the promoters of the megaproject aim to destabilise the narrative of improper stakeholder consultation to create a counter-narrative of effective consultation.

The protesters also claim that the megaproject affects people from whom land is taken and also people living near the project. While people who are affected by land acquisition are paid compensation, those who are only near to the project receive nothing. The protesters also strive to create a narrative that the project will cause pollution and other construction hassles, resulting in decreasing property values for people living near the project.

3.3.3 Environmental sustainability

Another aspect of evolving narratives in the megaproject is in the area of environmental sustainability. One of the benefits of going for new transportation projects such as a high-speed rail network or a metro rail project is because it is an environmentally friendly alternative of travel in contrast to private transport or fossil fuel-dependent public modes of transport. Promoters of such projects emphasise that the project is a low-carbon and environmentally sustainable transport solution to create a favourable narrative of the project. However, the protesters argue that environmental sustainability also includes the construction phase and the concerned megaproject is not the most sustainable option. They highlighted how trees are cut down along the project to enable the construction of the megaproject, which is not at all environmentally sustainable. The promoters of the megaproject oppose the environmental sustainability counter-narrative by claiming that the people opposing the environmental sustainability of the megaproject have vested interests against it. They highlight that people living near the megaproject want to stop the megaproject coming up in their backyard, i.e., Not In My Backyard (NIMBY), and are focusing on environmental sustainability for their cause. NIMBY is a common term used for a group of people who protest against an endeavour because it is harmful to their own neighbourhood (Lake, 1993). Thus, the narrative of environmental sustainability of a megaproject also evolves through multiple rounds of counter-narratives.

3.3.4 Excessive noise

Excessive noise from the megaproject can also be one of the points of concern for protesters. They strive hard to create a narrative that the project is a disturbance to those people living nearby. On the other hand, there is a need for the promoters to create a narrative that the noise level is tolerable and within the acceptable limits. To achieve this, the promoter can compare the noise from the trains with the noise from highways and claim that the occasional noise from trains is more acceptable than the constant buzz, both day and night, from the highways. In this manner, the promoters aim to create a narrative that the noise level is comparable to those in other infrastructure projects such as highways. The promoters also compare the noise level of the megaprojects with similar projects in other countries that are already operational to claim that there are no complaints regarding noise for those projects.

3.4 Interaction between the narratives of promoters and protesters

The interaction between the narratives of promoters and protesters occurs through multiple iterations, such as setting up a narrative, setting up a counter-narrative, and countering the counter-narrative. These are discussed below.

3.4.1 Setting up a narrative

Narratives can be set up by the promoters or the protesters of a megaproject. For example, the promoters set up the narrative of the need for the megaproject and the narrative of environmental sustainability, while the protesters set up the narrative of the stakeholder consultation process and the narrative of excessive noise, as discussed above. Setting up a narrative can important in endorsing the vested interest of the stakeholder group. The promoters want to create a favourable image of the megaproject and create a narrative in support of the megaproject, as in the case of the narrative of the need for the megaproject and the narrative of environment sustainability. The protesters, in contrast, want to topple the megaproject and hence created a narrative against the project, such as with the narrative of the stakeholder consultation process and the narratives of excessive noise. People of standing play a key role in setting up a narrative through trust transference (Lim et al., 2006). This might occur, for example, with the prime minister or the transport secretary setting up the narrative of the need for the megaproject, or a well-connected member of the protestors' group highlighting the narrative of the stakeholder consultation process. Narratives that are set up by people of reputation are covered by the media and gain traction (Hung, 2014). These narratives are subsequently either built upon or countered.

3.4.2 Setting up a counter-narrative

A counter-narrative contests the narrative which is set up by the promoters or promoters of the megaproject. The narrative of the need for the megaproject set up by the promoters of the megaproject are countered by the protesters of the megaproject. Similarly, the narrative of the stakeholder consultation process set up by the protesters of the megaproject are countered by the promoters of the megaproject. Again, people of reputation play an important role in setting up the counter-narrative as the only narratives backed with referent power (French & Raven, 1959; Ninan et al., 2019) is taken up by the media and reported. As discussed, counter-narratives are dependent on the dominant narrative and are referenced and positioned against the dominant narrative (McLean, 2015). Counter-narratives challenge dominant narratives, yet they can also be challenged and changed by other counter-narratives (Frandsen et al., 2017). In project settings, even the narrative over the Amsterdam metro project as a major technological innovation was challenged by the counter-narratives of citizens emphasising the damage done to the old city (Van den Ende & Van Marrewijk, 2019). Hence, there is an ongoing process of multiple negotiations and contestations of meanings, values, identities, and images, fought through multiple battles of narratives in megaproject settings.

3.4.3 Countering the counter-narrative

The counter-narrative is also subject to contesting by the promoters and protesters of the project. We can trace multiple rounds of the interaction between narratives and counter-narratives in the case of narrative of the need for the megaproject, the narrative of the stakeholder consultation process, the narrative of environmental sustainability, or the narrative of excessive noise, in different projects. Narratives are unstable and can shift from one equilibrium to another, as seen in the work of Todorov (1971). Sorsa and Vaara (2020) observe that as different stakeholders advance their arguments based on the stance of the other, they appropriate elements from each other's arguments, and thereby arrive at a convergence of arguments. Thus, as Van Marrewijk (2017) record, megaprojects embedded in the society are multivocal, can change over time, and can be strategic in power struggles.

3.5 Contesting the narratives

Resisting the counter-narrative involves the processes followed in handling the pull to create a counter-narrative by another stakeholder group. The organisation that seeks to stabilise the narrative when a counter-narratives emerge resort to different strategies. The strategies involved in resisting the counter-narrative include rejecting, delaying, and accepting part of the narrative, each of which is discussed below.

3.5.1 Rejecting

The organisation seeking to stabilise the narrative can reject the counter-narrative and argue that the counter-narrative does not hold. They give evidence for rejecting the counter-narrative and even highlight why the counter-narrative emerged. One of the frequent narratives stressed by the promoters of the megaproject is that the megaproject is environmentally friendly. The protesters, through highlighting that the project would damage an area of outstanding natural beauty or put endangered species at risk, create a counter-narrative claim that the megaproject is not environmentally friendly. The counter-narrative highlights that the megaproject is not good for the environmental landscape of the country. To resist this counter-narrative, the promoters of the megaproject argues that the project will not go through that particular area or that is will not create as much environmental damage as claimed by the protesters, and thereby dismissing the counter-narrative. Thus, the promoters of the project can reject the counter-narrative of the project being not environmentally sustainable through simple reasoning. Along with rejecting the counter-narrative, the promoters often seek to destabilise the credibility of the protesters by claiming that they are NIMBYs and that they always oppose the project by drawing on wider arguments against the project. Rejecting the counter-narrative giving proper reasons helps the narrative to continue and not be affected by the pull of the counter-narrative.

3.5.2 Delaying

Another strategy to handle the pull to create a counter-narrative is to delay the counter-narrative. For example, if the protesters claim that the project passes through a cemetery and that dead bodies would need to be exhumed to make way for the project, in an attempt to create a counter-narrative that there are problems with the current route, the promoters try to delay action. In a similar situation in the HS2 project in the UK, the project spokesperson delayed the counter-narrative from being dominant by saying that the project will look into the concern (Hough, 2012). In the process, the project team destabilises further discussion on the topic by acknowledging the concerns of the protesters regarding exhuming dead bodies are real, and thereby buys time to take a final decision. By employing this strategy, promoters postpone the decision on this topic and do not let the counter-narrative topple the narrative of the need for project. Similarly, in other projects when different environmental issues such as cutting down the oldest wild pear tree in the country, and social issues such as moving a place of worship with significant associated religious sentiment to make way for the project, the promoter of the project resists the counter-narrative by claiming that they are investigating the issue and adequate actions will be taken subsequently. The strategy adopted here is to delay the counter-narrative and thereby prevent it from destabilising the dominant narrative of the need for project. This strategy is similar to the term the 'political long grass' (Hood et al., 2007), by which tricky issues are made the subject of a long inquiry, thereby pushing them down the news agenda.

3.5.3 Accepting

When either rejecting and delaying of the counter-narrative is not an option, the promoter will accept part of the narrative. It should be noted that public infrastructure projects should engage stakeholders and construct with minimum disruptions to the community and maximum value for the society. The rejecting and delaying strategy is only recommended for noisy stakeholders, who might not be the most affected stakeholder, but try to get maximum vested interest from the project either by changing parts of the project or by cancelling the project altogether. Multiple infrastructure projects accept part of the narrative of the legitimate stakeholder and made amendments to it. As highlighted earlier, one of the most common promoters' narratives is of the project being environmentally friendly. To destabilise this narrative in the case of the HS2 project in the UK, the protesters claimed that the project was harming the Chilterns' ecologically sensitive area as a counter-narrative. The project team accepted part of this counter-narrative and made amendments to the route by adding 7.5 miles of tunnelling and 3.5 miles of deep cuttings along the 13 miles of proposed line through the Chilterns. Similarly, when the protesters claimed that the people who are affected by land acquisition are paid compensation while those who are near to the project get no money, the project agreed to buy and lease back homes affected by the route to destabilise this counter-narrative. Therefore, the project accepted part of the counter-narrative, finding aspects of their analysis and their demands to be legitimate, thereby mitigating some of the negative effects of the pull to create a counter-narrative.

Figure 3.1 shows the three strategies employed to resist the counter-narrative discussed above. In the case of rejecting and delaying, the existing narrative continues; by contrast, in the case of accepting part of the counter-narrative, the narrative also evolves such that the project is not wholly cancelled. Sorsa and Vaara (2020) highlight that narratives progress through a process of struggles, ambiguity and contradictions. It is seen from the case study that narratives experience a pull to create a counter-narrative from those opposing the narrative. The most preferred strategy

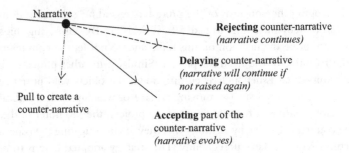

FIGURE 3.1 Strategies employed to resist the counter-narrative

to resist the counter-narrative is to reject it by showing proper evidence for rejection and even destabilising the credibility of those creating the counter-narrative. By rejecting the counter-narrative, the narrative continues unaffected. Another strategy to resist the counter-narrative is to delay the counter-narrative through the above-mentioned 'political long grass'. Over time, the narrative will continue unaffected if the counter-narrative is not raised again. Additionally, the supporter of the narrative can accept part of the counter-narrative, thereby reducing the impact of the pull of the counter-narrative. In the process, the narrative evolves. Through these three strategies, the narrative of the project evolves in the process of the interaction between narratives and counter-narratives. Aaltonen and Sivonen (2009) discuss how projects respond to stakeholder pressures through adaptations, compromises, avoidances or dismissals. Similarly, negotiations, trade-offs, incentives and concessions (Chinyio & Akintoye, 2008; Yang et al., 2014) can be employed to manage stakeholders' demands. Adding to the existing literature, this research highlights similar processes for handling and thereby managing counter-narratives.

3.6 Conclusion

In this chapter we describe the different stakeholders in megaprojects. We highlight promoters as stakeholders interested in supporting the completion of a megaproject, such as government agencies, contractors and lobby groups. We highlight protesters as stakeholders interested in derailing the megaproject, such as people affected by the project, activists, and non-governmental organisations. The conflict between promoters and protesters are common in megaprojects and we describe how narratives are key to resolving these conflicts. We describe different promoters' and protesters' narratives involved in the need for the project, the stakeholder consultation process, environmental sustainability and excessive noise. We also highlight how the interaction between narratives exist in practice with multiple steps such as setting up a narrative, setting up a counter-narrative, and countering the counter-narrative. The different ways in which the narrative can be contested such as rejecting, delaying and accepting are also discussed.

CASE STUDY 3: MELBOURNE EAST WEST LINK, AUSTRALIA

The 2008 report by Sir Rod Eddington, titled 'East West Link Needs Assessment,' warned about the steady port freight and population growth in Melbourne, Australia and how it is rapidly leading to the congestion of the city's roads. The report highlighted the growing west-to-east travel demand and the resulting imbalance between population growth and employment opportunities in the city's western suburbs and eastern suburbs. Eddington stated that a link between the east and the west of the city would relieve congestion on Hoddle Street, reduce east–west rat running through suburban streets, assist north–south traffic flows including public transport, and improve accessibility to city jobs for the residents of the western suburbs. It would enhance the connectivity between the port and freeway network and thereby encourage more trucks on to the appropriate freeway network. It would be intended to carry as much as 20 per cent of passengers, workers and freight between Melbourne Airport and the eastern suburbs, thereby developing the area. The report studied the need for more east–west transport corridors within Melbourne and recommended an east–west road link between the Eastern Freeway, the Citylink, and possibly the Western Ring Road.

The arguments for the road were that the freeway would help relieve the traffic on the West Gate Bridge and M1 corridor, and that it would enhance connectivity between the freeway network and the Port of Melbourne. The project was seen as a top-priority infrastructure requirement for Melbourne. The Liberal party government, under the Premier of Victoria, planned to proceed with the East West link project just prior to the November 2014 state election. Thus, it was proposed that an East West Link, an 18-kilometre highway in Melbourne, Australia, would be built to connect the Eastern Freeway at Clifton Hill with the Western Ring Road at Sunshine West. The project's first stage was planned to be a 4.4-km tunnel from Hoddle Street in Clifton Hill to CityLink at Parkville. The government signed a $5.3-billion contract with the East West Connect consortium in September 2014 for the completion of the stage by early 2020. The second stage, which is the western section between Parkville and the Western Ring Road, was expected to commence in late 2015 and be completed by 2023. The project's total cost was estimated at between $15 and $17 billion. At this cost, the project was the most expensive road project in Australia's history and an indicative toll of $5.50 was proposed for cars.

The East West Connect consortium started preparation works for the project. The residents of Collingwood were told that their houses might be compulsorily acquired, and regular protests were seen at the drilling sites for the project because of the land acquisition. As time progressed, the project begun to attract major controversy within the public, with questions over its effectiveness in terms of reducing congestion, its prioritising cars over public transport, the transparency of business case and the local effect on parks and zoos. Critics claimed that the need for the project was not supported by traffic statistics and that the Victoria state government should instead have used funds for public transport, namely the Melbourne Metro Rail Project. There were also complaints that it was a misuse of public funds, and it would fail to alleviate congestion and would lead to the permanent loss of parts of the Royal Park. Local councils and public transport advocates opposed the project, and several community groups were formed to block its construction. Thus, the East West Link project encountered strident opposition from a range of stakeholders, among them academics, architects, resident action groups, industry groups, local councils, developer lobbies, the Greens, with some strongly opposed to the project in its entirety. The growing sentiments with the public were that the link would result in more private vehicles on the road, that most motorists on the Eastern Freeway were trying to get into the Central Business District (CBD) rather than going around it, and that more money should instead be spent on public transport. A protest in the project is shown in Figure 3.2.

At this point the opposition Labour party leader Daniel Andrews announced his intentions to scrap the entire $15–17 billion East West Link project if he were to come into power at the Victoria state election in November 2014. As the race to the state election began, the infrastructure project became the

FIGURE 3.2 Protests in the Melbourne East West link project

most contentious point of the entire election campaign, with two clear sides: The ruling Liberal party for the project and the opposition Labour party against it. This was confirmed by the then Prime Minister of Australia, Tony Abbott, who dubbed the 2014 Victoria state election a referendum on the East West Link project. Andrews remained determined to cancel the entire road even when it was clear that great sums of compensation would be required if the project was scrapped as contracts had already been signed and the construction of the road was almost underway.

In the aftermath, the Labour party won the election with 47 seats, with the Liberal party on 38 seats. It marked the first time in over 60 years that a party in Victoria had only been in office for one term. The following day Daniel Andrews confirmed that the East West Link would be cancelled. Subsequently, the project was abandoned by the newly elected Labour party government in 2015, with Andrew at the helm. The cost of the project's cancellation was $1.1 billion, in the form of penalties and compensations. The people of Victoria state had to shell out half a billion to not build a road.

After the cancellation of the project, the federal government has refused to allow the allotted $4 billion for the link to be spent on any other Victorian project. The poor connectivity between Melbourne's Eastern Freeway and CityLink continues to be stressed by Infrastructure Australia. The 2019 Infrastructure Australia audit found that the east–west corridor to the north of Melbourne's CBD had the highest road congestion delay cost in Melbourne. Hence, the project still features in the list of Australia's 32 highest-priority infrastructure needs. The project has been recommended by many studies and has been identified as a key need for decades. Therefore, supporters of the scheme believe the project will be built in future; however, the later the construction the higher the cost.

Exercise

1. Identify the promoter and the protester narratives of the project.
2. Why was the protester narrative successful?
3. What can be done to make the promoter narrative successful?

Sources

ABC news. (2021). Victorian opposition renews East West Link push ahead of next state election. https://www.abc.net.au/news/2021-04-15/victorian-liberals-revive-east-west-link-plan/100070446 (Accessed on 15 August 2022).

Edwards, J. (2015). East West Link: Cost of scrapping project more than $1.1 billion, auditor-general says. https://www.abc.net.au/news/2015-12-09/auditor-general-reports-on-east-west-link-costs/7012618 (Accessed on 15 August 2022).

Roads Australia. (2018). East West Link: Did Melbourne need it? https://roadsaustralia. weebly.com/east-west-link.html (Accessed on 15 August 2022).

References

Aaltonen, K., & Sivonen, R. (2009). Response strategies to stakeholder pressures in global projects. *International Journal of Project Management*, 27(2), 131–141.

Abolafia, M. Y. (2010). Narrative construction as sensemaking: How a central bank thinks. *Organization Studies*, 31(3), 349–367.

Achterkamp, M. C., & Vos, J. F. (2008). Investigating the use of the stakeholder notion in project management literature, a meta-analysis. *International Journal of Project Management*, 26(7), 749–757.

Bartel, C. A., & Garud, R. (2009). The role of narratives in sustaining organizational innovation. *Organization Science*, 20(1), 107–117.

Bekdik, B., & Thuesen, C. (2016). Reinventing the hospital – A study of lost synergies in Danish healthcare. In *Engineering Project Organization Conference Proceedings*. Washington, USA.

Biesenthal, C., Clegg, S., Mahalingam, A., & Sankaran, S. (2018). Applying institutional theories to managing megaprojects. *International Journal of Project Management*, 36(1), 43–54.

Black, C., Akintoye, A., & Fitzgerald, E. (2000). An analysis of success factors and benefits of partnering in construction. *International Journal of Project Management*, 18(6), 423–434.

Boddy, D., & Paton, R. (2004). Responding to competing narratives: Lessons for project managers. *International Journal of Project Management*, 22(3), 225–233.

Bourne, L., & Walker, D. H. T. (2005). Visualising and mapping stakeholder influence. *Management Decision*, 43(5), 649–660.

Chinyio, E. A., & Akintoye, A. (2008). Practical approaches for engaging stakeholders: Findings from the UK. *Construction Management and Economics*, 26(6), 591–599.

Cooren, F. (1999). Applying socio-semiotics to organizational communication: A new approach. *Management Communication Quarterly*, 13, 294–304.

Dailey, S. L., & Browning, L. (2014). Retelling stories in organizations: Understanding the functions of narrative repetition. *Academy of Management Review*, 39(1), 22–43.

Datta, A., Ninan, J., & Sankaran, S., (2020). 4D visualization to bridge the knowing-doing gap in megaprojects: An Australian case study. *Construction Economics and Building*, 20(4), 25–41.

Derakhshan, R., Turner, R., & Mancini, M. (2019). Project governance and stakeholders: A literature review. *International Journal of Project Management*, 37(1), 98–116.

Di Maddaloni, F., & Davis, K. (2017). The influence of local community stakeholders in megaprojects: Rethinking their inclusiveness to improve project performance. *International Journal of Project Management*, 35(8), 1537–1556.

Donaldson, T., & Preston, L. E. (1995). The stakeholder theory of the corporation: Concepts, evidence, and implications. *Academy of Management Review*, 20, 65–91.

Duman, D. U., Green, S. D., & Larsen, G. D. (2018). Historical narratives as strategic resources: Analysis of the Turkish international contracting sector. *Construction Management and Economics*, 36(1), 1–17.

Dunham, L., Freeman, R.E., & Liedtka, J. (2006). Enhancing stakeholder practice: A particularized exploration of community. *Business Ethics Quarterly*, 16(1), 23–42.

Enninga, T., & van der Lugt, R. (2016). The innovation journey and the skipper of the raft: About the role of narratives in innovation project leadership. *Project Management Journal*, 47(2), 103–114.

Frandsen, S., Kuhn, T., & Lundholt, W. (2017). *Counter-narratives and organization*. Routledge: New York and London.

Flyvbjerg, B. (1998). *Rationality and power: Democracy in practice*. Chicago: University of Chicago Press.

Flyvbjerg, B. (2014). What you should know about megaprojects and why: An overview. *Project Management Journal*, 45(2), 6–19.

Flyvbjerg, B., Bruzelius, N., & Rothengatter, W. (2003). *Megaprojects and risk: An anatomy of ambition*. Cambridge, UK: Cambridge University Press.

Freeman, R. E. (1984). *Strategic management: A stakeholder approach*. Boston, MA: Pitman.

French, J. R. P., Jr., & Raven, B. H. (1959). The bases of social power. In D. Cartwright (ed.), *Studies in social power* (pp. 150–167). Ann Arbor, MI: Institute for Social Research.

Ganz, M. (2011). Public narrative, collective action, and power. In S. Odugbemi & T. Lee (eds.), *Accountability through public opinion: From inertia to public action*, (pp. 273–289). Washington: The World Bank.

Garud, R., & Turunen, M. (2017). The banality of organizational innovations: Embracing the substance-process duality. *Innovation*, 19(1), 31–38.

Grayson, K. (1997). Special session summary narrative theory and consumer research: Theoretical and methodological perspectives. In M. Brucks & D. J. MacInnis (eds.), *NA – Advances in consumer research* (Vol. 24, pp. 67–70). Provo, UT: Association for Consumer Research.

Havermans, L. A., Keegan, A., & Den Hartog, D. N. (2015). Choosing your words carefully: Leaders' narratives of complex emergent problem resolution. *International Journal of Project Management*, 33(5), 973–984.

Henisz, W. J., Dorobantu, S., & Nartey, L. J. (2014). Spinning gold: The financial returns to stakeholder engagement. *Strategic Management Journal*, 35(12), 1727–1748.

Hood, C., Jennings, W., Hogwood, B., & Beeston, C. (2007). *Fighting fires in testing times: exploring a staged response hypothesis for blame management in two exam fiasco cases*. London: London School of Economics.

Hough, A. (2012). HS2: 50,000 bodies 'to be exhumed' for high speed rail link. *The Telegraph*. https://www.telegraph.co.uk/news/uknews/road-and-rail-transport/9189588/HS2-50000-bodies-to-be-exhumed-for-high-speed-rail-link.html (accessed on 10 October 2022).

Hung, K. (2014). Why celebrity sells: A dual entertainment path model of brand endorsement. *Journal of Advertising*, 43(2), 155–166.

Lake, R. W. (1993). Planners' alchemy transforming NIMBY to YIMBY: Rethinking NIMBY. *Journal of the American Planning Association*, 59 (1), 87–93.

Li, B., Akintoye, A., Edwards, P. J., & Hardcastle, C. (2005). Critical success factors for PPP/PFI projects in the UK construction industry. *Construction Management and Economics*, 23(5), 459–471.

Li, T. H., Ng, S. T., & Skitmore, M. (2012). Conflict or consensus: An investigation of stakeholder concerns during the participation process of major infrastructure and construction projects in Hong Kong. *Habitat International*, 36(2), 333–342.

Lim, K. H., Sia, C. L., Lee, M. K., & Benbasat, I. (2006). Do I trust you online, and if so, will I buy? An empirical study of two trust-building strategies. *Journal of Management Information Systems*, 23(2), 233–266.

Lindgren, M., & Packendorff, J. (2007). Performing arts and the art of performing – On co-construction of project work and professional identities in theatres. *International Journal of Project Management*, 25(4), 354–364.

Littau, P., Jujagiri, N. J., & Adlbrecht, G. (2010). 25 years of stakeholder theory in project management literature (1984–2009). *Project Management Journal*, 41(4), 17–29.

Liu, B., Li, Y., Xue, B., Li, Q., Zou, P. X. W., & Li, L. (2018). Why do individuals engage in collective actions against major construction projects? An empirical analysis based on Chinese data. *International Journal of Project Management*, 36(4), 612–626.

Lounsbury, M., & Glynn, M. A. (2001). Cultural entrepreneurship: Stories, legitimacy, and the acquisition of resources. *Strategic Management Journal*, 22, 545–564.

McLean, K. C. (2015). *The co-authored self: Family stories and the construction of personal identity*. New York, NY: Oxford University Press.

Miller, R., Lessard, D. R., & Sakhrani, V. (2017). Megaprojects as games of innovation. In B. Flyvbjerg (ed.), *The Oxford handbook of megaproject management* (pp. 217–237). Oxford: Oxford University Press.

Mitchell, R. K., Agle, B. R., & Wood, D. J. (1997). Toward a theory of stakeholder identification and salience: Defining the principle of who and what really counts. *Academy of Management Review*, 22(4), 853–886.

Mok, K. Y., Shen, G. Q., & Yang, J. (2015). Stakeholder management studies in mega construction projects: A review and future directions. *International Journal of Project Management*, 33(2), 446–457.

Ninan, J., Mahalingam, A., & Clegg, S., (2019). External stakeholder management strategies and resources in megaprojects: An organizational power perspective. *Project Management Journal*, 50(1), 625–640.

Ninan, J., Mahalingam, A., Clegg, S., & Sankaran, S. (2020). ICT for external stakeholder management: Sociomateriality from a power perspective. *Construction Management and Economics*, 38(9), 840–855.

Ninan, J., & Sergeeva, N. (2021). Labyrinth of labels: Narrative constructions of promoters and protesters in megaprojects. *International Journal of Project Management*, 39(5), 496–506.

Ninan, J., Mahalingam, A., & Clegg, S. (2022). Asset creation team rationalities and strategic discourses: Evidence from India. *Infrastructure Asset Management*, 8(2), 1–10.

Noland, J., & Phillips, R. (2010). Stakeholder engagement, discourse ethics and strategic management. *International Journal of Management Reviews*, 12(1), 39–49.

Oppong, G. D., Chan, A. P., & Dansoh, A. (2017). A review of stakeholder management performance attributes in construction projects. *International Journal of Project Management*, 35(6), 1037–1051.

Orr, R. J., & Scott, W. R. (2008). Institutional exceptions on global projects: A process model. *Journal of International Business Studies*, 39 (4), 562–588.

Parmar, B. L., Freeman, R. E., Harrison, J. S., Wicks, A. C., Purnell, L., & De Colle, S. (2010). Stakeholder theory: The state of the art. *Academy of Management Annals*, 4(1), 403–445.

Polkinghorne, D. E. (1991). Narrative and self-concept. *Journal of Narrative and Life History*, 1(2), 135–153.

Rappaport, J. (2000). Community narratives: Tales of terror and joy. *American Journal of Community Psychology*, 28(1), 1–24.

Ricoeur, P. (1991). Life in quest of narrative. In D. Wood (ed.), *On Paul Ricoeur: Narrative and interpretation* (pp. 20–33). New York, NY: Routledge.

Roper, S., & Davies, G. (2007). The corporate brand: Dealing with multiple stakeholders. *Journal of Marketing Management*, 23, 75–90.

Sergeeva, N. (2019). *Making sense of innovation in the built environment*. Oxfordshire, UK: Routledge.

Sergeeva, N., & Winch, G. M. (2021). Project narratives that potentially perform and change the future. *Project Management Journal*, 52(3), 264–277.

Smith, J., & Love, P. E. (2004). Stakeholder management during project inception: Strategic needs analysis. *Journal of Architectural Engineering*, 10(1), 22–33.

Sorsa, V., & Vaara, E. (2020). How can pluralistic organizations proceed with strategic change? A processual account of rhetorical contestation, convergence, and partial agreement in a Nordic City Organization. *Organization Science*, 31(4), 839–864.

Todorov, P. (1971). *The poetics of prose.* Oxford: Blackwell.

Turner, J. R., Lecoeuvre, L., Sankaran, S., & Er, M. (2019). Marketing for the project: Project marketing by the contractor. *International Journal of Managing Projects in Business,* 12(1), 211–227.

Vaara, E., Sonenshein, S., & Boje, D. (2016). Narratives as sources of stability and change in organizations: Approaches and directions for future research. *Academy of Management Annals,* 10(1), 495–560.

Van den Ende, L., & Van Marrewijk, A. H. (2019). Teargas, taboo and transformation: A neo-institutional study of public resistance and the struggle for legitimacy in an Amsterdam subway project. *International Journal of Project Management,* 37(2), 331–346.

Van Marrewijk, A. (2007). Managing project culture: The case of Environ Megaproject. *International Journal of Project Management,* 25(3), 290–299.

Van Marrewijk, A. (2017). The multivocality of symbols: A longitudinal study of the symbolic dimensions of the high-speed train megaproject (1995–2015). *Project Management Journal,* 48(6), 47–59.

Van Marrewijk, A., Clegg, S. R., Pitsis, T. S., & Veenswijk, M. (2008). Managing public-private megaprojects: Paradoxes, complexity, and project design. *International Journal of Project Management,* 26(6), 591–600.

Veenswijk, M., & Berendse, M. (2008). Constructing new working practices through project narratives. *International Journal of Project Organisation and Management,* 1(1), 65–85.

Viitanen, K. Falkenbach, H., & Nuuja, K. (2010). *Compulsory purchase and compensation – recommendations for good practice.* Copenhagen, Denmark: International Federation of Surveyors.

Vrijhoef, R., & Koskela, L. (2000). The four roles of supply chain management in construction. *European Journal of Purchasing & Supply Management,* 6(3-4), 169–178.

Winch, G. M. (2007). Managing project stakeholders. In P. Morris & J. Pinto (eds.), *The Wiley guide to project, program, and portfolio management* (pp. 271–289). Hoboken, NJ: John Wiley & Sons.

Yang, R. J., Wang, Y., & Jin, X.-H. (2014). Stakeholders' attributes, behaviors, and decision-making strategies in construction projects: Importance and correlations in practice. *Project Management Journal,* 45(3), 74–90.

Zakhem, A. (2007). Stakeholder management capability: A discourse-theoretical approach. *Journal of Business Ethics,* 79(4), 395–405.

4

CRAFTING NARRATIVES IN MEGAPROJECTS

4.1 Introduction

Narratives can be crafted or produced in many ways as part of discourses and communication (Vaara et al., 2016). For instance, Sergeeva (2017) record how narratives are crafted through labels and how labels are used meaningfully and purposefully in organisations. There are other discursive ways in which narratives are represented in organisations, such as in the form of verbal, visual or written forms, or a combinations of all three (Sergeeva, 2019). Narrative tools such as these can persuade people to change, get them to work together, enable knowledge transfer, neutralise the rumour mill and create a compelling new future (Denning, 2005). In this chapter, we aim to describe the different discursive instruments that can be used to craft narratives in megaprojects.

4.2 Narrative instruments

We highlight various narrative instruments, such as labels, stories and comparisons, and record how they are effective in crafting narrative in megaprojects.

4.2.1 Labels

The Webster dictionary defines a label as 'a word or phrase that identifies something or someone'. Labels are quasi-objects that easily travel and translate ideas from one place to another (Czarniaswka & Joerges, 1995). It directs attention and helps audiences perceive an emerging pattern (Kennedy, 2008). Existing research claims that labels shape identity along with defining memberships and bestowing legitimacy to the organisation. Sergeeva (2017) notes that identity is linguistically indexed through labels. Ewing (2004) records that when an agent attaches a label

DOI: 10.1201/9781003248378-4

to the subject position they have occupied, the label becomes an identity. Abrams (2003) notes that both identity and labels are mutually reinforcing. Vergne and Wry (2014) record that labels are crucial when organisations seek membership into an existing category. Labels have significant effects on the functioning of markets and help people make sense of organisation and their activities (Porac et al., 2011). Logue and Clegg (2015) highlight that labels may work to legitimise (or illegitimise) organisations and practices. Because of the importance of labels, they are treated as a resource in organisations (Granqvist et al., 2013) and can be called as a 'technology of control' (Suchman, 1994). Labels are also an instrument of power through which the relationships between class interests and institutional processes are constructed and sustained (Wood, 1989). Labels are even used as marketing tools and have marketing implications (Benders et al., 1998). Douglas (1986) had long back called for more studies on the role of labels in organisation studies as they are used to build, reinforce, and reflect broader systems of value, meaning and power.

Labelling is he process of giving a name to an emerging phenomena based on a common set of attributes and roles (Negro et al., 2011). It may be carried out for a person, place, group, event or any other key element in a narrative (Baker, 2006). In labelling, organisations use the cultural, material and social resources to influence the perceptions resulting from the label (Slavich et al., 2020). On the strategic intent of labelling, Sergeeva (2017) notes that some labels are used meaningfully, while others are used purposefully to achieve a goal, such as to shape the identity of an organisation as 'innovative'. These labels can be either maintained or contested. When labels are maintained, everyone in the organisation understands the label, its relevance and benefits (Sergeeva, 2014). Labels are also contested, which leads to inconsistencies (Vergne & Swain, 2017). Controversies in a label can destabilise an organisation and hamper its survival (Slavich et al., 2019). The strategic intent of labelling also involves the manipulation of existing labels, thereby resulting in the stigmatised identity of the competitors being diluted (Vergne, 2012). Logue and Clegg (2015) studied how organisations make sense of events by exploring the practices and politics of labelling. Thus, the labelling process is of significant importance for crafting megaprojects narrative and thereby the process of megaproject organising and disorganising.

Van Marrewijk (2007) notes how employees are labelled as the 'Gideon's gang' – a biblical metaphor for a brave group of men that knows no fear and uses creative, innovative methods to reach their goals. As a result of this labelling, the Environ megaproject employees considered in the work by Van Marrewijk (2007) were able to strongly identify themselves with the organisation's innovative and entrepreneurial culture. Labels used as part of creating a megaproject narrative can shape and sustain the individual's and organisation's identity, such as being perceived as 'innovative' or 'largest' (Ninan et al., 2022; Sergeeva, 2014; Sergeeva & Zanello, 2018). Labels are used for different purposes in megaprojects and they highlight various elements such as its 'novelty' and 'originality' (Sergeeva, 2017). Within megaproject settings, labels such as largest, sustainable and efficient are frequently exercised and there is currently a dearth in megaproject management literature on how these labels come into becoming, are contested, and are maintained.

Label can be assigned for the project, people or practices (Ninan & Sergeeva, 2021). While the promoters label the megaproject favourably, using terms such as 'fast' and 'low-carbon', the protesters label it as a 'vanity project' and as a project 'for the rich', thereby seeking to destabilise the project. Labels such as 'fast' and 'low-carbon' seek to make the megaproject attractive while labels such as 'vanity' and 'for the rich' strive to make the megaproject unappealing. In all cases, labels change the identity of the megaproject and influence the interpretation of the organisation (Czarniawska-Joerges, 1994). Labelling at the organisational level seeks to modify the identity, image and reputation of the megaproject organisation (Ashforth & Humphrey, 1997). While positive labels such as being 'innovative' and 'learning' can improve the acceptability of the megaproject organisation (Sergeeva, 2017; Sergeeva & Roehrich, 2018; Sergeeva & Zanello, 2018), negative labels such as 'regime' (which implies an authoritarian kind of ruling) have a negative connotation to the organisation (Shahi & Talebinejad, 2014).

In the case of labels of people, the promoters of the megaproject often label a section of the protesters as NIMBYs, a common label used for a person or organisation that object to an endeavour because of its harmful effects on their own neighbourhood (Lake, 1993). In contrast, the protesters call themselves 'friends of the earth', identifying themselves as fighting for the Earth and against the environmental damage the megaproject is causing. Labels such as 'NIMBY' or 'friends of the earth' create personal identities to the groups labelled. While promoters label themselves as 'railway enthusiasts,' protesters refer to them as 'railway lunatics' (Ninan & Sergeeva, 2021). Eyben and Moncrieffe's (2006) record the use of label 'citizen' for claiming an identity card which dictates resources, status and services. Moncrieffe and Eyben (2013) highlight that the way people are labelled matters as it has a long-standing influence on how the groups perceive themselves, respond to opportunities, make claims and exercise agency. So, positive labels such as 'friends of the earth' for protesters and 'railway enthusiasts' for promoters can awaken pride and help in organising, while negative labels, such as 'NIMBY' for protesters and 'railway lunatics' for promoters, can shame people and result in disorganising.

In the case of labels of practices, the promoters label the stakeholder consultation practice 'transparent', while the protesters label it a 'farce'. In other cases, the selection process of the project can be labelled as 'very impressive' by the promoters and as a 'betrayal' by the protesters; similarly, the compensation scheme can be labelled as 'unprecedented' by the promoters and as 'sinister' by the protesters (Ninan & Sergeeva, 2021). The labelling of an issue or practice as a threat or an opportunity affects subsequent information processing and even the motivations of the key decision-makers (Dutton & Jackson, 1987). Cohen (2007) notes that the labelling of a practice as 'routine' seeks to guarantee that the execution of choices will be pretty much automatic. In the case of labelling the consultation practice as 'transparent' or a 'farce', the proponents of the label try to highlight that the label 'is' or 'will be' the practice.

As highlighted earlier, the role of labels in the sensemaking process, the organisations involved or the process through which these labels are maintained and contested are not explored in an organisational setting. The conflicting objectives of the promoters and the protesters in a megaproject setting and the prevalent use of labels in them provides us an opportunity to explore this theoretical gap. It was seen that both promoters and protesters employed labels to market themselves in situations where there was conflicting interests and identity crisis for giving direction and guiding sensemaking (Ancona, 2011). While project management literature discusses favourable labels such as 'novelty', 'innovative' and 'largest' (Ninan et al., 2022; Sergeeva, 2017; Sergeeva & Roehrich, 2018), the literature falls short of discussing unfavourable labels. We note that while the promoters assigned favourable labels such as 'fast' or 'low-carbon' to the project, the protesters assigned unfavourable labels such as 'vanity' or 'for the rich'. Contrary to the findings of Negro et al. (2010), that organisations whose identity spans multiple labels tend to be less legitimate, here we see the promoters and the protestors creating multiple identities for the megaproject by employing different labels. Our findings back the contributions of Pontikes (2012), who highlight that organisations in emerging markets with multiple and ambiguous labels are like chameleons as they are able to respond easily to changes. We argue that megaprojects are similar to organisations in emerging markets and use multiple labels to adapt easily to changes in the megaproject environment.

Project management literature highlights marketing as one aspect of stakeholder management and discusses the project, convincing the stakeholders that the benefits from the project are greater than the losses (Turner & Lecoeuvre, 2017; Turner et al., 2019). However, marketing is not exclusive to the promoters of the megaproject, as protesters also use marketing of their interests to disrupt the project. Protesters' organisations such as interest groups and action groups are also involved in the marketing activities of megaprojects such as the project itself, investors and contractors, albeit from a different interest. It should also be noted that the focus of project management has shifted from the project manager to the whole of the project team (Crawford, 2000) and hence project marketing should be performed by all members of the organisation (Turner & Lecoeuvre, 2017).

Wood (1989) records that labels of people such as 'poor' are often used in development policy. However, in the case of megaprojects labels were assigned to the megaproject, people and practice in order to shape the meaning of the megaproject, the people associated with it, or the practices followed. This is similar to the findings of Baker (2018), who found that labelling is carried out for a person, place, group, event or any other key element in a narrative. As highlighted earlier, labels of people assign an identity card which dictates resources, status and services (Eyben & Moncrieffe, 2006). Such use of labels for people is used to discipline and cast specific attributions of identity to people (Foucault, 1977). The use of labels such as 'NIMBY', as discussed earlier, can be a means of moral disciplining (Logue & Clegg, 2015). In contrast, labels of megaproject change the identity of the megaproject and influence the interpretation of the organisation

(Czarniawska-Joerges, 1994). Labels of megaprojects, such as 'for the rich' in this instance, can shape the rules of meaning and can have implications for decisions regarding the megaproject. Labels of practice are dependent on the identity labels of the megaproject and people and can result in different treatments from those to whom the label is not applied (Dix et al., 2020). For example, the label of consultation practice as a 'farce' or 'transparent' can be explained through the label of megaproject as 'for the rich' or the label of people as 'NIMBY', respectively. If the megaproject is identified as 'for the rich' and as 'unnecessary', then the consultation practice will be seen overtly as a 'farce', while if the identity of the protesting people are seen as 'NIMBY', then the consultation practice will be seen overtly as 'transparent'. Thus, the labels of practice only make sense in the context of labels of people and labels of the megaproject. It should be noted that the labels of practice can also affects subsequent information processing and even the motivations of key decision-makers (Dutton & Jackson, 1987) by shaping the labels of the megaproject and people in a circular loop.

4.2.2 Stories

Stories are personalised, entertaining and emotional in nature (Vaara et al., 2016) and can generate a common understanding and shared vision amongst organisational members (Perkins et al., 2017). Boje (2008) records that narratives are mobilised through storytelling and records that there is a 'story turn' before the 'narrative turn'. Storytelling is defined as an activity of telling or sharing stories about personal experiences, life events and situations (Sergeeva & Trifilova, 2018). Stories are way more than the presentation of information or facts; rather, they re-present facts in an elegant way (Gabriel, 2004). Stories exist in organisations as fully developed stories with a beginning and an end or in fragments as bits and pieces (Boje, 1991). They configure apparently independent and disconnected elements of existence into related parts of a whole (Barry & Elmes, 1997). By presenting information as part of a complex whole, organisational stories are more easily learned and remembered by others (Shaw et al., 1998). They are part of the organisational discourse and can construct identities and interests across space and time (Vaara & Tienari, 2011).

Grayson (1997) highlights that stories have, for a long time, been tools of persuasion, stretching as far back as Aesop's fables and right up to the present day, with TV shows such as Sesame Street. Stories also entertain, explain, inspire, educate, convince, generate and sustain meaning (Gabriel, 2000). They are conceived by the sender with an intention to convey a meaning to an audience (Pace, 2008). Personal accounts or stories by farmers losing the value of their land can create a narrative that the project has negative social consequences. Even accounts or stories by people displaced by the project seek to create empathy and understanding from others (Gabriel, 2000). Such stories help individuals understand and describe who they are. By telling stories about their experiences, they play leading roles in the stories and construct themselves as characters whose attributes can be revealed and communicated (Gergen, 1999). Thus, narratives shape not only how individuals view themselves, but also how others view them (O'Connor, 2004). When these stories are reported and circulated in social media, news articles and other avenues, it can create a narrative that the community does not need the project.

Gabriel (2000) notes that stories are not just descriptions but also an avenue for emotional engagement with the audience. Stories are personalised, entertaining and emotional in nature (Sergeeva & Green, 2019; Vaara et al., 2016). The stories of the people displaced by the project lead readers to empathise with these stories and create a shared vision that the government is forcing the compulsory acquisition of properties that stand in their way. This shared vision can result in a call for organising among people to fight to keep their countryside. Thus, as Weick et al. (2005) claim stories can be considered as being an integral part of organising. Stories help in organising as it generates a common understanding and shared vision amongst members of either promoter or protester groups in the case of megaprojects (Perkins et al., 2017; Sarpong & Maclean, 2012).

4.2.3 Comparisons

Comparisons involve comparing oneself with another. They are a daily activity and help people make sense of information (O'Neill et al., 2007). There are many cues in everyday life, such as informative statements, opinions, jokes, etc. (Seligman, 2006). Lamertz (2002) notes comparison as an important social cue and therefore as a key source of sensemaking. Salancik and Pfeffer (1978) note that judgements are developed subsequent to comparison with others. People give importance to comparisons, even when they have adequate information that they are doing better than average (Seta et al., 2006). Different types of comparisons are discussed in the literature. Comparisons with self are known as self-referent comparisons, whereas comparisons with others are called other-referent comparisons. Self-referent comparisons in the form of comparisons of performance with expectations are instrumental for fulfilment (O'Neill & Mone, 2005). Other-referent comparisons in the form of comparisons with the treatment of peers help in determining fairness (Lamertz, 2002).

People's comparisons with one another are discussed at length in the literature surrounding social comparisons theory (Greenberg et al., 2007). Suls et al. (2002) define social comparison as comparing oneself with others in order to evaluate or to enhance some aspects of the self. Festinger (1954), in his theory of social comparison, suggests that people evaluate information sources in terms of personal relevance, using similar others for comparison. He notes that the more similar someone is, the more relevant his or her views are for understanding one's own world. After all, meanings are 'relational and comparative' as meaning derives in part from comparisons between categories (Tajfel & Turner, 1986). Social comparisons can be either planned or emergent. Planned comparisons involve a supervisor evaluating the performance of a subordinate relative to others (Mumford, 1983).

In contrast, emergent comparisons are naturally occurring instances, such as when, for example, an employee compares their own pay with that of others (Blysma & Major, 1994). There can also be comparisons with those who are doing better called 'upward comparisons' and with those who are doing worse called 'downward comparisons' (Buunk et al., 1990).

Greenberg et al. (2007) discuss the role of comparisons in organisational justice, performance appraisal, virtual work environments, affective behaviour in the workplace and even leadership. The social comparison also gives an idea of fairness in the eyes of the beholder (Lamertz, 2002). Festinger (1954) noted that the more ambiguous the situations, the more people rely on comparisons to assess them. He noted that even in situations of uncertainty individuals seek comparative information. Therefore, comparisons are important for organisations in their initial stages as this period is the one most marked with ambiguousness and uncertainty.

Within the megaproject context, comparisons with context exist in the form of comparisons with the economic context, governance context and the state of transportation context. Comparisons with economic context involves the promoter-initiated comparisons of comparing the economy of a country lagging behind those of other countries, thereby necessitating the completion of the megaproject. Comparisons with context can also be employed by the protesters of the megaproject. They can use comparisons to highlight that the country with the proposed megaproject does not have sufficient money to invest in such a project, in contrast to the other countries. Comparisons with the governance context can also be seen in megaprojects. Protesters can claim that the governance process in the country, such as the selection of the megaproject and the business case of the megaproject, does not stack up in comparison to how businesses are awarded in a popular television show *Dragon's Den* (Ninan & Sergeeva, 2022). The protesters also often compare the governance of the project and compulsory land acquisition process with historic times such as the civil war when democracy was suspended. History has a role in the making and the unmaking of organisational order (Hansen, 2012) and such use of history for purposes in the present is called the 'uses of the past' (Wadhwani et al., 2018). Another type of comparison with context is the comparison with the state of transportation context. Here a comparison is made with the state of infrastructure or transportation that exists in the country or elsewhere and not with any project per se. Such comparisons are used by protesters' groups to claim that the country does not need this type of project because of its transportation context, such as there being no long distance to cover.

In megaprojects, there are also comparisons with other projects within the country, with projects outside the country and even with the project itself. Comparisons with projects within the country involves comparisons with the performance and practices of other megaprojects. For example, the protesters of the megaproject warn promoters that protests in this megaproject would be worse than protests in earlier megaprojects. Zelditch et al. (1970) record that local comparisons alone are insufficient to promote feelings of inequality. Added to these, both the promoter and protester employ comparisons with the megaproject itself. For example, protesters such as the opposition party claim that if they were in power, they would have started the construction of the project earlier than the current government's plan.

Thus, comparisons are seen as an important instrument to craft a narrative in megaprojects by both the promoters and the protesters. Similar to organisations, social identities are created in megaprojects by intergroup comparisons where group members seek to enhance their self-esteem through drawing distinctions between their groups and others (Ashforth & Mael, 1989).

4.3 Narrative instruments and their functions

The narrative instruments of stories, labels and comparisons serve different functions in megaproject organising. While stories help in creating a shared vision of the project, labels help in creating an identity for the project, and comparisons help in enhancing the perception of justice for the community as shown in Figure 4.1.

Taken together, these different functions contribute to a public image for the project which can lead to either external stakeholder support for or resistance to the project (Oppong et al., 2017). The perception of justice can lead to external stakeholder acceptance or rejection of the project and the project purpose (El-Sawalhi & Hammad, 2015). Whether positive or negative, these community experiences with the megaproject organisation can influence its legitimacy in the eyes of the external stakeholders (Derakhshan et al., 2019).

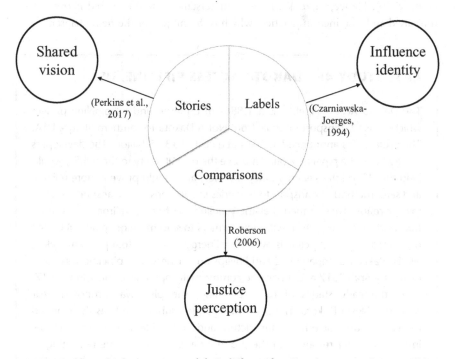

FIGURE 4.1 Narrative instruments and their different functions in megaproject organising

4.4 Conclusion

In this chapter, we describe the different instruments to craft a narrative. Labels are words or phrases assigned to someone that is instrumental in forming their identity and labels such as largest, sustainable, efficient, etc., are frequently exercised in megaproject settings. Stories are personalised, entertaining and emotional in nature and they are instrumental in creating a shared vision amongst organisational members. Comparisons involve comparing oneself with another and they prove influential in how people make sense of information. We highlight how the narrative instruments of stories, labels and comparisons serve different functions in project organising. While labels help in creating an identity for the project, stories help in creating a shared vision of the project, and comparisons help in enhancing the perception of justice for the community.

Even though there are isolated instances of storytelling, labels and repetition within organisational settings, there is still a lack of clear understanding as to how these help in creating a narrative. Riessman (2002) stresses the importance of tools and structures employed by the narrator and calls for more research to uncover them. Similarly, Sergeeva (2019) highlights that multiple narratives exist in an organisational setting as a 'meshwork' of discourses and activities. In addition to the spoken and written narrative instruments discussed in this chapter, there can also be visual instruments such as a glossy picture of how the project would look when completed, which can bring stakeholders together to complete the project (Datta et al., 2020). Hence, there is a need to understand how the crafted narratives are mobilised and sustained in practice, which is the subject of the next chapter.

CASE STUDY 4: DAKOTA ACCESS PIPELINE, USA

The Dakota Access Pipeline is a 1,886-km underground oil pipeline project which aims to transport oil shale from North Dakota to southern Illinois, USA. The project was announced in 2014 at a cost of $3.78 billion. The developers claimed that the pipeline would improve the overall safety to the public, would help the US to attain energy independence, and would prove a more reliable and safer method of transport to refineries in comparison to transport through rails or roads. They argued that the pipeline will free up railroads and other transportation links, which will allow farmers to ship more grain and generate more income. The pipeline is owned by Energy Transfer (36.4 per cent stake), MarEn Bakken Company LLC and Phillips 66 Partners. The pipeline was completed by April 2017 and it became commercially operational on 1 June 2017.

In the early stages of route planning, the pipe was proposed near Bismarck, North Dakota. The plan was changed subsequently as the planned route was not near other infrastructure and in consideration of its potential impact on water resources in the proposed area. However, the re-routing of the pipeline from Bismarck, which has 92 per cent white population, to the

Standing Rock Sioux Reservation, a native American region, was termed 'environmental racism' and 'state-sanctioned racial violence'. Conservation groups were worried about the safety of the project, and the impacts on air, water, wildlife and farming, because of the risk of pipeline disruption. They claimed that an oil spill would threaten both their water supply and their cultural resources. Native Americans accused the government of hastily approving each stage of the review process and ignoring federal regulations and established treaties with Native American tribes. They protested how there was a complete lack of environmental foresight and consideration in the project proposal. The Standing Rock Sioux Tribe claim that the project violates Article II of the Fort Laramie Treaty, which guarantees the "undisturbed use and occupation" of land they hold. The Tribe passed a resolution regarding the pipeline stating that "the Dakota Access Pipeline poses a serious risk to the very survival of our Tribe and ... would destroy valuable cultural resources."

Protests against the Dakota Access Pipeline occurred at several places because of concerns about the pipeline's impact on the environment and on sites sacred to Native Americans. Iowa farmers filed lawsuits to prevent the state from using this eminent domain to take their land. In North Dakota, next to and on the Standing Rock Indian Reservation, nearly 15,000 people from around the world protested, staging a sit-in which lasted several months. Protest were in the form of runs, horseback rides and marches. Many celebrities, politicians and native allies supported the movement and joined the protest. Figure 4.2 shows the protests in the project wherein the protesters urge the government to honour the treaty rights for the native Americans.

FIGURE 4.2 Protests against the Dakota Access Pipeline

Protests at pipeline construction sites in North Dakota began in the spring of 2016. It drew Indigenous people, who called themselves 'water protectors' and 'land defenders', from across North America as well as many other supporters, creating the largest gathering of Native Americans in the past hundred years. In July, 'ReZpect Our Water', a group of Native American youth, ran from Standing Rock in North Dakota to Washington, DC to raise awareness of what they perceive as a threat to their people's drinking water and that of everyone who relies on the Missouri and Mississippi rivers for drinking water and irrigation.

The government called the pipeline opposition movement 'an ideologically driven insurgency with a strong religious component' that operated along a 'jihadist insurgency model'. The protesters were cast as economic terrorists and saboteurs when in fact they were going out and having their voices heard about why these pipelines are problematic for their communities and the environment.

The police cleared protesters' camps using long-range acoustic devices, tear gas and stun bags. Some protesters had built barricades; they attached themselves to construction equipment for the Dakota Access Pipeline, and started vandalising property. On one occasion, when the protesters tried to stop bulldozers from excavating part of the pipeline's route which could damage Native American cemeteries, security officers deployed at the site attacked them with pepper spray and attack dogs. The protesters called this a "provocation of peaceful resistance to violence". The activists also faced a slew of lawsuits, which they call an orchestrated campaign to target environmentalists and indigenous peoples.

In March 2020, a court ordered the United States Army Corps of Engineers to conduct a new environmental impact assessment because the government had not conducted an adequate study of the pipeline's effects on the community. In a subsequent judgement, the pipeline was ordered to be closed and the oil removed from it, pending a new environmental impact assessment and its clearance. After an appeal, during which they assured a new environmental impact assessment would be carried out, the project avoided the temporary closure and assured a new environmental impact assessment. The pipeline is still in use.

Exercise

1. What are the promoters' and protesters' narratives in the Dakota Access Pipeline project?
2. What are the different labels assigned to the protesters?

Sources

Brown, A., Parrish, W., & Speri, A., (2017). Leaked documents reveal counterterrorism tactics used at standing rock to "defeat pipeline insurgencies". https://theintercept.com/2017/05/27/leaked-documents-reveal-security-firms-counter terrorism-tactics-at-standing-rock-to-defeat-pipeline-insurgencies/ (Accessed on 18 August 2022).

Cagle, S. (2019). 'Protesters as terrorists': Growing number of states turn antipipeline activism into a crime. https://www.theguardian.com/environment/2019/jul/08/wave-of-new-laws-aim-to-stifle-anti-pipeline-protests-activists-say (Accessed on 18 August 2022).

CBC News. (2016). Norway's biggest bank may reconsider Dakota Access funding. https://www.cbc.ca/news/business/dakota-access-bank-dnb-1.3841739 (Accessed on 18 August 2022).

Frazin, R. (2020). Court cancels shutdown of Dakota Access Pipeline. https://thehill.com/policy/energy-environment/510748-court-cancels-shutdown-of-dakota-access-pipeline/ (Accessed on 18 August 2022).

Gilio-Whitaker, D. (2019). *As long as grass grows: The indigenous fight for environmental justice, from colonization to Standing Rock*. Boston, MA.

MacPherson, J., & Nicholson, B. (2016). Armed police begin operation to clear North Dakota pipeline protest camp. https://www.thestar.com/news/world/2016/10/27/armed-police-begin-operation-to-clear-north-dakota-pipeline-protest-camp.html (Accessed on 18 August 2022).

Petroski, W. (2016). 296 Iowa landowners decline Bakken pipeline. https://eu.desmoinesregister.com/story/news/local/government/2016/02/09/296-iowa-land owners-decline-bakken-pipeline/80085982/ (Accessed on 18 August 2022).

References

Abrams, L. S. (2003). Contextual variations in young women's gender identity negotiations. *Psychology of Women Quarterly*, 27(1), 64–74.

Ancona, D. (2011). Sensemaking: Framing and acting in the unknown. In S. Snook, N. Nohria, & R. Khurana (eds.), *The handbook for teaching leadership* (pp. 3–19). Thousand Oaks, CA: Sage Publications.

Ashforth, B. E., & Humphrey, R. H. (1997). The ubiquity and potency of labeling in organizations. *Organization Science*, 8(1), 43–58.

Ashforth, B. E., & Mael, F. (1989). Social identity theory and the organization. *Academy of Management Review*, 14: 20–39.

Baker, S. M. (2006). Consumer normalcy: Understanding the value of shopping through narratives of consumers with visual impairments. *Journal of Retailing*, 82(1), 37–50.

Baker, M. (2018). *Translation and conflict: A narrative account*. New York, NY: Routledge.

Barry, D., & Elmes, M. (1997). Strategy retold: Toward a narrative view of strategic discourse. *Academy of Management Review*, 22: 429–452.

Benders, J., van den Berg, R. J., & van Bijsterveld, M. (1998). Hitch-hiking on a hype: Dutch consultants engineering re-engineering. *Journal of Organizational Change Management*, 11(3), 201–215.

Blysma, W. H., & Major, B. (1994). Social comparisons and contentment: Exploring the psychological costs of the gender wage gap. *Psychology of Women Quarterly*, 18, 241–249.

Boje, D. M. (1991). The storytelling organization: A study of story performance in an office-supply firm. *Administrative Science Quarterly*, 36(1), 78–108.

Boje, D. (2008). *Storytelling organizations*. London: Sage.

Buunk, B. P., Collins, R. L., Taylor, S. E., Van Yperen, N. W., & Dakof, G. A. (1990). The affective consequences of social comparison: Either direction has its ups and downs. *Journal of Personality and Social Psychology*, 59, 1238–1249.

Cohen, M. D. (2007). Reading Dewey: Reflections on the study of routine. *Organization Studies*, 28(5), 773–786.

Crawford, L. (2000). Project management competence for the new millennium. In *Proceedings of 15th World Congress on Project Management*, IPMA, London, England.

Czarniawska-Joerges, B. (1994). Narratives of individual and organizational identities. *Communication Yearbook*, 17, 193–221.

Czarniaswka, B., & Joerges, B. (1995) Winds of organizational change: How ideas translate into objects and actions. In S. Bacharach, P. Gagliardi, & B. Mundell (eds.), *Research in the sociology of organizations* (pp. 171–209). Greenwich, NJ: JAI Press.

Datta, A., Ninan, J., & Sankaran, S., (2020). 4D visualization to bridge the knowing-doing gap in megaprojects: An Australian case study. *Construction Economics and Building*, 20(4), 25–41.

Derakhshan, R., Turner, R., & Mancini, M. (2019). Project governance and stakeholders: A literature review. *International Journal of Project Management*, 37(1), 98–116.

Denning, S. (2005). Transformational innovation: A journey by narrative. *Strategy & Leadership*, 33(3), 11–16.

Dix, N., Lail, A., Birnbaum, M., & Paris, J. (2020). Exploring the "At-Risk" student label through the perspectives of higher education professionals. *Qualitative Report*, 25(11), 3830–3846.

Douglas, M. (1986). *How institutions think*. Syracuse, NY: Syracuse University Press.

Dutton, J. E., & Jackson, S. E. (1987). Categorizing strategic issues: Links to organizational action. *Academy of Management Review*, 12(1), 76–90.

El-Sawalhi, N. I., & Hammad, S. (2015). Factors affecting stakeholder management in construction projects in the Gaza Strip. *International Journal of Construction Management*, 15(2), 157–169.

Ewing, K. (2004). Migration, identity negotiation and self-experience. In J. Friedman, & S. Randeria (eds.), *Worlds on the move: Globalization, migration and cultural security* (pp. 117–140). New York: I. B. Tauris & Co. Ltd.

Eyben, R., & Moncrieffe, J. (2006). *The power of labelling in development practice*. Institute of Development Studies.

Festinger, L. (1954). A theory of social comparison processes. *Human Relations*, 7(2), 117–140.

Foucault, M. (1977). *Discipline and punish: The birth of the prison*. London, England: Allen & Lane.

Gabriel, Y. (2000). *Storytelling in organizations: Facts, fictions, and fantasies*. Oxford, UK: Oxford University Press.

Gabriel, Y. (2004). Narratives, stories and texts. In D. Grant, C. Hardy, C. Oswick, & L. Putnam (eds.), *The Sage handbook of organizational discourse* (pp. 61–77). London: Sage.

Gergen, K. J. (1999). *An invitation to social construction*. Thousand Oaks, CA: Sage.

Granqvist, N., Grodal, S., & Woolley, J. L. (2013). Hedging your bets: Explaining executives' market labeling strategies in nanotechnology. *Organization Science*, 24(2), 395–413.

Grayson, K. (1997). Special session summary narrative theory and consumer research: Theoretical and methodological perspectives. In M. Brucks & D. J. MacInnis (eds.),

NA – Advances in consumer research (Vol. 24, pp. 67–70). Provo, UT: Association for Consumer Research.

Greenberg, J., Ashton-James, C. E., & Ashkanasy, N. M. (2007). Social comparison processes in organizations. *Organizational Behavior and Human Decision Processes*, 102(1), 22–41.

Hansen, P. H. (2012). Business history: A cultural and narrative approach. *Business History Review*, 86(4), 693–717.

Kennedy, M. T. (2008). Getting counted: Markets, media, and reality. *American Sociological Review*, 73, 270–295.

Lake, R. W. (1993). Planners' alchemy transforming NIMBY to YIMBY: TRthinking NIMBY. *Journal of the American Planning Association*, 59 (1), 87–93.

Lamertz, K. (2002). The social construction of fairness: Social influence and sense making in organizations. *Journal of Organizational Behavior*, 23(1), 19–37.

Logue, D. M., & Clegg, S. R. (2015). Wikileaks and The News of the World: The political circuitry of labeling. *Journal of Management Inquiry*, 24(4), 394–404.

Moncrieffe, J., & Eyben, R. (2013). *The power of labelling: How people are categorized and why it matters*. London, UK: Earthscan.

Mumford, M. D. (1983). Social comparison theory and the evaluation of peer evaluations: A review and some applied implications. *Personnel Psychology*, 36, 867–881.

Negro, G., Hannan, M. T., & Rao, H. (2010). Categorical contrast and audience appeal: Niche width and critical success in winemaking. *Industrial and Corporate Change*, 19, 1397–1425.

Negro, G., Hannan, M. T., & Rao, H. (2011). Category reinterpretation and defection: Modernism and tradition in Italian winemaking. *Organization Science*, 22(6), 1449–1463.

Ninan, J., Mahalingam, A., & Clegg, S. (2022). Power in news media: Framing strategies and effects in infrastructure projects. *International Journal of Project Management*, 40(1), 28–39.

Ninan, J., & Sergeeva, N. (2021). Labyrinth of labels: Narrative constructions of promoters and protesters in megaprojects. *International Journal of Project Management*, 39(5), 496–506.

Ninan, J., & Sergeeva, N. (2022). Mobilizing megaproject narratives for external stakeholders: A study of narrative instruments and processes. *Project Management Journal*, 53(5), 520–540.

O'Connor, E. (2004). Storytelling to be real: Narrative, legitimacy building and venturing. In D. Hjorth & C. Steyaert (eds.), *Narrative and discursive approaches in entrepreneurship* (pp. 105–124). Northampton, MA: Elgar.

O'Neill, B. S., Halbesleben, J. R., & Edwards, J. C. (2007). Integrating employment contracts and comparisons: What one can teach us about the other. *Journal of Managerial Issues*, 19(2), 161–185.

O'Neill, B. S., & Mone, M. A. (2005). Psychological influences on referent choice. *Journal of Managerial Issues*, 17(3), 273–292.

Oppong, G. D., Chan, A. P., & Dansoh, A. (2017). A review of stakeholder management performance attributes in construction projects. *International Journal of Project Management*, 35(6), 1037–1051.

Pace, S. (2008). YouTube: An opportunity for consumer narrative analysis? *Qualitative Market Research: An International Journal*, 11(2), 213–226.

Perkins, G., Lean, J., & Newbery, R. (2017). The role of organizational vision in guiding idea generation within SME contexts. *Creativity and Innovation Management*, 26(1), 75–90.

Pontikes, E. G. (2012). Two sides of the same coin: How ambiguous classification affects multiple audiences' evaluations. *Administrative Science Quarterly*, 57, 81–118.

Porac, J. F., Thomas, H., & Baden-Fuller, C. (2011). Competitive groups as cognitive communities: The case of Scottish Knitwear manufacturers revisited. *Journal of Management Studies*, 48, 646–664.

Riessman, C. K. (2002). Narrative analysis. In A. M. Huberman & M. B. Miles (eds.), *The qualitative researcher's companion* (pp. 17–70). Thousand Oaks, CA: Sage.

Salancik, G. R., & Pfeffer, J. (1978). A social information processing approach to job attitudes and task design. *Administrative Science Quarterly*, 224–253.

Sarpong, D., & Maclean, M. (2012). Mobilising different versions for new product innovation. *Technovation*, 32(12), 694–702.

Seligman, L. (2006). Sensemaking throughout adoption and the innovation-decision process. *European Journal of Innovation Management*, 9(1), 108–120.

Sergeeva, N. (2014). Understanding of labelling and sustaining of innovation in construction: A sensemaking perspective. *Engineering Project Organization Journal*, 4(1), 31–43.

Sergeeva, N. (2017). Labeling projects as innovative: A social identity theory. *Project Management Journal*, 48(1), 51–64.

Sergeeva, N., & Green, S. D. (2019). Managerial identity work in action: Performative narratives and anecdotal stories of innovation. *Construction Management and Economics*, 37(10), 604–623.

Sergeeva, N., & Roehrich, J. (2018). Learning in temporary multi-organizations: Constructing identities to realize performance improvements. *Industrial Marketing Management*, 75, 184–192.

Sergeeva, N., & Trifilova, A. (2018). The role of storytelling in innovation process. *Creativity and Innovation Management*, 27(4), 1–10.

Sergeeva, N., & Zanello, C. (2018). Championing and promoting innovation in UK megaprojects. *International Journal of Project Management*, 36(8), 1068–1081.

Sergeeva, N. (2019). *Making sense of innovation in the built environment*. Oxfordshire, UK: Routledge.

Seta, J. J., Seta, C. E., & McElroy, T. (2006). Better than better-than average (or not): Elevated and depressed self-evaluations following unfavorable social comparisons. *Self and Identity*, 5, 51–72.

Shahi, M., & Talebinejad, M. R. (2014). Frame labelling of competing narratives in journalistic translation. *Research in Applied Linguistics*, 5(2), 23–40.

Shaw, G., Brown, R., & Bromiley, P. (1998). Strategic stories: How 3M is rewriting business planning. *Harvard Business Review*, 76(3), 41–50.

Slavich, B., Svejenova, S., Opazo, M. P., & Patriotta, G. (2020). Politics of meaning in categorizing innovation: How chefs advanced molecular gastronomy by resisting the label. *Organization Studies* 41(2), 267–290.

Suchman, L. (1994). Do categories have politics? *Computer Supported Cooperative Work*, 2, 177–190.

Suls, J., Martin, R., & Wheeler, L. (2002). Social comparison: Why, with whom, and with what effect?. *Current Directions in Psychological Science*, 11(5), 159–163.

Tajfel, H., & Turner, J. C. (1986). The social identity theory of intergroup behavior. In S. Worchel & W. G. Austin (eds.), *Psychology of intergroup relations* (2nd ed., pp. 7–24). Chicago, IL: NelsonHall.

Turner, J. R., & Lecoeuvre, L. (2017). Marketing by, for and of the project: Project marketing by three types of organizations. *International Journal of Managing Projects in Business*, 10(4), 841–855.

Turner, J. R., Lecoeuvre, L., Sankaran, S., & Er, M. (2019). Marketing for the project: Project marketing by the contractor. International Journal of Managing Projects in Business, 12(1), 211–227.

Vaara, E., Sonenshein, S., & Boje, D. (2016). Narratives as sources of stability and change in organizations: Approaches and directions for future research. *Academy of Management Annals*, 10(1), 495–560.

Vaara, E., & Tienari, J. (2011). On the narrative construction of multinational corporations: An antenarrative analysis of legitimation and resistance in a cross-border merger. *Organization Science*, 22(2), 370–390.

Van Marrewijk, A. (2007). Managing project culture: The case of Environ Megaproject. *International Journal of Project Management*, 25(3), 290–299.

Vergne, J. P. (2012). Stigmatized categories and public disapproval of organizations: A mixed-methods study of the global arms industry, 1996–2007. *Academy of Management Journal*, 55, 1027–52.

Vergne, J. P., & Swain, G. (2017). Categorical anarchy in the UK? The British media's classification of bitcoin and the limits of categorization. *Research in the Sociology of Organizations*, 51, 187–222.

Vergne, J. P., & Wry, T. (2014). Categorizing categorization research: Review, integration, and future directions. *Journal of Management Studies*, 51(1), 56–94.

Wadhwani, R. D., Suddaby, R., Mordhorst, M., & Popp, A. (2018). History as organizing: Uses of the past in organization studies. *Organization Studies*, 39(12), 1663–1683.

Weick, K. E., Sutcliffe, K. M., & Obstfield, D. (2005). Organizing and the process of sensemaking. *Organization Science*, 16(4), 409–421.

Wood, G. (1989). *Labelling in development policy*. London: Sage.

Zelditch, M., Berger, J., Anderson, B., & Cohen, B. P. (1970). Equitable comparisons. *Pacific Sociological Review*, 13(1), 19–26.

5

MOBILISING NARRATIVES IN MEGAPROJECTS

5.1 Introduction

Even though there are isolated instances of storytelling, labels and repetition within organisational settings, there is still a lack of understanding around how these help in creating a narrative. Riessman (2002) stresses the importance of tools and structures employed by the narrator and calls for more research to uncover them. Similarly, Sergeeva (2019) highlights that multiple narratives exist in an organisational setting as a 'meshwork' of discourses and activities.

Along with discourses, there are also activities and processes that help in mobilising narratives. Notable among these is the idea of narrative repetition advanced in the work of Dailey and Browning (2014). They record how stories are repeated in organisations whether over the water cooler or in a formal quarterly meeting, yet researchers give little attention to the form, function and implications of the recurrence of stories. They note that retelling is an important component of narrative theory as it performs functions such as control/resistance, integration/differentiation and stability/change within organisations.

Here, we differentiate between narrative instruments and narrative processes to understand how narratives are mobilised in megaprojects. In our work we define narrative instruments as tools, devices or resources which enable the creation of a narrative such as stories, labels and comparisons as discussed in the previous chapter. We define narrative processes as the technique, methods or procedure that is followed for using instruments such as repeating, endorsing, humourising and actioning; these are discussed in this chapter. Thus, narrative instruments are used in particular narrative processes to achieve a purpose, i.e., mobilising the project narrative. To understand how narratives are mobilised in a megaproject setting, we need to explore the narrative instruments and narrative processes together.

DOI: 10.1201/9781003248378-5

5.2 Narrative processes

There are different processes through which the crafted narratives can be mobilised and sustained as there are counter-narratives seeking to disrupt them. Among the processes to mobilise narratives are repeating, endorsing, humourising and actioning.

5.2.1 Repeating

Repeating involves the retelling of narratives in organisations and serves important functions such as control/resistance, integration/differentiation and stability/change, as noted in the work of Dailey and Browning (2014). They record how stories are repeated in organisations, whether over the water cooler or in a formal quarterly meeting.

The different narrative instruments, such as stories, labels and comparisons, that try to create a narrative that the community does not need the project are often repeated so that it becomes stable. Garud and Turunen (2017) note that retelling stories is a way of reinforcing cultural norms and values. Kotter (2012) highlight that ideas sink in only after they have been heard many times. Repeating is largely discussed in the literature for stories (Dailey & Browning, 2014; Garud & Turunen, 2017). In the case of megaprojects, labels and comparisons were also repeated in the process of mobilising narratives. Sometimes narratives are repeated through the use of synonyms, acronyms or by simply explaining the narrative (Ninan & Sergeeva, 2021). News articles covering the project can be an important avenue to repeat the narratives of the project (Ninan et al., 2022). Love and Ahiaga-Dagbui (2018) highlight that the persistent reverberation of the convenient narratives of optimism bias and strategic misrepresentation in both academia and media as the causes of megaproject failures has led to these explanations becoming an accepted norm.

5.2.2 Endorsing

Support from people who occupy a prominent status in the society can help in mobilising a narrative. Endorsement from elected representatives, academicians, celebrities, etc., can give narratives authenticity. Elected representatives such as the prime minister of a country hold significant influence in the society and their use of comparisons to signify that the project would help the country compete with other countries can be significant in mobilising narratives. The narrative that the community does not need the project are sometimes supported by academicians who have expertise in the area of economics and social development. Within megaprojects, publicising visits by regional leaders and celebrities to the construction site is discussed in Ninan et al. (2019) as a branding strategy which can prove effective in changing the project community to advocates of the project. Endorsing of the stories, labels or comparisons by people who occupy eminent positions can enhance trust. Lim et al. (2006)

record that trust transference through associations with existing reputed people or organisations is instrumental in trust-building. Thus, the endorsement of narratives by prominent people results in the community trusting the megaproject almost as much as they trust the prominent person/organisation.

5.2.3 Humourising

Humourising is another process through which the crafted narratives can be mobilised. This involves the use of different forms of comic engagements to make narratives circulate more and reach wider parts of the society. Instances within projects that have humour in them were shared and enjoyed by the community, and these can help in the process of mobilizing a narrative. For instance, in the case of the HS2 megaproject, the protesters created a parody video[1] entitled the *Downfall of HS2*, dubbing a famous scene from *Downfall* – a film that charts Adolf Hitler's final days in his Berlin bunker. The video was created in an attempt to shape a narrative on the flawed business case. A news article reported on the video that was widely shared as below,

> The HS2 parody, which was put online this week, starts with Hitler – in the role of Secretary of State for Transport – saying to his generals: "Don't worry, we have the business case" … However, one of his staff replies: "It barely breaks even, despite the insane increase in traffic we forecast." At this, Hitler flies into a rage: "Even with the million jobs we made up you still couldn't get it right. We said it would reduce flights from Heathrow – it won't. It'll only create 10,000 jobs in the Midlands – less than last month's increase in unemployment in Birmingham alone."
>
> *(Quoted from the news article 'HS2 rail link gets Hitler parody',*
> *dated 4 October 2012)*

The video used humour in comparing the secretary of state for transport to Hitler. The need for the project was labelled as an 'insane increase in traffic' and the whole parody was structured as a form of storytelling, being made personalised, entertaining and emotional in nature (Vaara et al., 2016). Similarly, the protesters used humour by comparing the quick travel which the project would enable with common daily durations such as ordering food and watching a game, as below,

> "London to Manchester in less time than it takes for United to play Arsenal? Birmingham to London quicker than it takes enjoy a pint and an order of fish and chips down the way at the Queen's Head pub? Birmingham to Leeds in the time it takes to enjoy a pot of tea? It could all happen – and via train, no less – now that the U.K. government has given the go-ahead to a national high speed rail network called HS2."
>
> *(Quoted from the news article 'UK high speed rail hs2*
> *gets go ahead' dated 20 January 2012)*

Jarzabkowski and Le (2017) record that humour can either affirm or shift an existing response in an organisation. Adding to the literature, we note that humour in the case of the HS2 megaproject was used to promote and market the discourses such as stories, comparisons, and labels in order to mobilise a narrative. Discourses that display a sense of humour are clearly memorable and more likely to be disseminated (Sergeeva & Green, 2019). Humour can also help to form a cohesive team, thereby bringing unity and also creating a positive cultural environment to help manage conflicts successfully (Ponton et al., 2020). Humourising to mobilise a narrative in projects can be seen in news media articles, public press meetings, and even in social media. In their analysis of 106,316 Facebook messages across 782 companies, Lee et al. (2018) found that the inclusion of humour can lead to greater consumer engagement. Users also have fun trolling the project in social media to highlight issues and poor performance during a project's operational phase (Mathur et al., 2021). Such trolling can change the community's perception of the project.

5.2.4 Actioning

Processes carried out to mobilise narratives include putting discourses into action. Here, narratives move from textual/spoken/visual form to an action form. These include activities such as walking along the entire project as a sign of resistance or protest, which is considered an activity to mobilise the narrative that the community does not need the project. Similarly, the promoters, such as the secretary of state for transport, also walked partly along the route in an attempt to reinforce the label of 'effective consultation process' (Jaspercopping, 2010). Grayson (1997) claim that narratives help us understand events. We extend this by highlighting that narratives and events are interrelated as events from the megaproject can be instrumental in mobilising narratives through actioning. Putting a discourse into action can help to embed the narrative in the mind of the community. As Weick (1988) notes, actions test the provisional understanding generated through prior sense-making and thereby strengthen existing narratives. Actions are a way we recognise important contributions, acknowledge a common identity, and deepen our sense of community (Ganz, 2011). Thus, the process of actioning in megaprojects reinforce the provisional understanding generated through narratives.

5.3 Crafting and mobilising for external stakeholders

Crafting and mobilising narratives has implications in terms of the external stakeholder management of megaprojects. Megaprojects across the world affect numerous external stakeholders as they create economic, political and environmental disruptions in society (Sturup, 2009). The vocal among these stakeholders' campaign against the project and try to achieve their vested interests through the project, thereby changing many features of the project (Flyvbjerg, 1998). The most affected stakeholder may not necessarily be the most vocal (Van Marrewijk et al., 2008) and hence the activities of the vocal few can result in the project not delivering on its intended benefits. Using an instrumental perspective of stakeholder management, we argue that megaproject narratives can be employed to manage external stakeholders. The discourses for stakeholder management from an instrumental perspective have a strategic intent with a focus on achieving an organisation's corporate objectives (Zakhem, 2007). This research highlights the instruments, processes and medium project's use to manage external stakeholders.

Research on narratives has emphasised the role of narrative tools or instruments such as stories (Boje, 2008) and labels (Granqvist et al., 2013), which can neutralise the rumour mill and create a compelling new future (Denning, 2005). We also highlight how stories, labels and comparisons are used as narrative instruments to create a project narrative. The narrative literature also highlights how processes, such as repeating narratives, can perform functions that stabilise or control the narrative (Dailey & Browning, 2014). We highlight how, along with repeating, other processes, such as endorsing, humourising and actioning, can help stabilise the narrative. We also contribute to theory by highlighting that all the narrative instruments such as stories, labels and comparisons are repeated, endorsed, humourised and actioned. The interaction between narrative instruments and processes (Figure 5.1) helps us understand how narratives are mobilised in practice for external stakeholder management. It should be noted that the model created does not differentiate between the order of use of narrative instruments such as

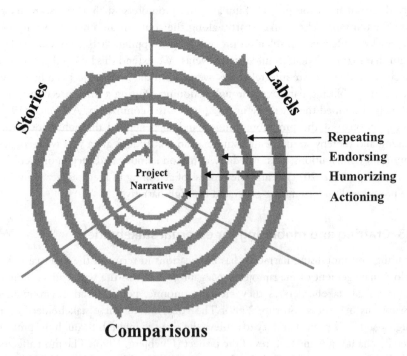

FIGURE 5.1 A simple model of the interaction between instruments and processes to mobilize project narratives

stories, labels, and comparisons, nor that of narrative instruments such as repeating, endorsing, humourising and actioning. Rather, the model only shows how a project narrative is created by different cycles of narrative instruments and processes. Future research can explore the role of counter-narratives (Andrews, 2002) or rhetorical contestations (Sorsa & Vaara, 2020) in the interaction between narrative instruments and processes.

We highlight news media as a medium for instrumental stakeholder management. In order to be resilient in the face of interest groups, it is important that projects have a good reputation and media image right from the start (Olander & Landin, 2008) and hence it is essential to build a favourable media narrative for the project. The narrative processes mobilise and bring people together. We also note that stories, labels and comparisons were individually or together repeated, endorsed, made attractive and actioned in order to build the narrative, as shown in Figure 5.2. Together, the narrative instruments and processes can help megaprojects mobilise a narrative that can potentially help in managing external stakeholders, such as gaining their acceptance and legitimacy.

We call on future studies to quantitatively explore the effectiveness of these instruments and processes towards achieving the project organisation's objectives. Practically, this research highlights the different ways in which narratives can be mobilised to improve external stakeholders' acceptance with regard to a proposed project or programme. Narrative instruments and processes have implications towards improving stakeholder acceptance through narratives. As stories can help create a shared vision of the project, the project team can bring about stories and the personal experiences of people who have benefited, or are projected to benefit, from the proposed megaproject. Such empathetic and real-life stories can be shared on social media, advertisements and project websites. As labels can help create a project identity, the project team can create and use labels such as 'largest consultation' or 'reliable service'. Adding to these, labels against the resistance group of the project can bring down their legitimacy and arguments. As comparisons can affect the perception of justice for the community, it is important that the project team should study the management of project-affected stakeholders in other projects and propagate the considerate practices adopted in their own project. As noted above, the resistance groups also

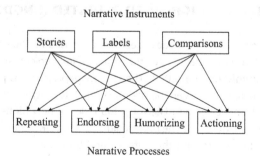

FIGURE 5.2 Instruments and processes for mobilizing narratives

leverage stories, labels and comparisons to create their favourable narrative. Hence it is important that these narrative instruments used by the project team are stabilised through repeating, endorsing, humourising and actioning. All of the narrative instruments can be repeated in news media, social media and other outlets by the project team. The project team should also reach out to leaders, celebrities and other people with referent power to endorse the different narrative instruments. Humourising and actioning of the narrative instruments can also help stabilise the project narratives. Recently, the project management literature has stressed the importance of marketing (Turner et al., 2019) and branding (Ninan et al., 2019) in project settings. The practice of marketing and branding strategies are prevalent in other sectors and project settings can adopt these to create and maintain a stable narrative. Such stable narratives can help projects to subtly create a 'reservoir of support' (Di Maddaloni & Davis, 2017) and thereby resistant the negative press and protests that seek to topple the project.

5.4 Conclusion

In this chapter, we describe the different narrative processes necessary to mobilise a narrative. Repeating involves the retelling of narratives in organisations and serves important functions such as control/resistance, integration/differentiation and stability/change within organisations. Endorsing involves getting support from people occupying a prominent status in the society and results in trust transference to the megaproject. Humourising involves the use of different forms of comic engagements to make narrative circulate more and reach wider parts of the society. Actioning involves moving narratives from textual/spoken/visual form to an action form and is instrumental in reinforcing the provisional understanding generated through narratives. Thus, in this chapter we highlight how along with repeating, other processes, such as endorsing, humourising and actioning, can help stabilise the narrative. We also note that stories, labels and comparisons were either individually or collectively repeated, endorsed, made attractive and actioned in order to build the narrative.

CASE STUDY 5:   HIGH SPEED 2, UNITED KINGDOM

The High-Speed Two (HS2) megaproject was proposed in 2009 as the second network of high-speed trains after the channel link project (renamed HS1), which was commissioned in 2003. The HS2 megaproject was planned to be delivered in multiple phases and aimed to connect the city centres of London, Birmingham, Manchester, and Leeds by 345 miles of new high-speed railway track as shown in Figure 5.3. The project aims to bring UK's cities closer to each other by effectively shrinking the distance and time taken to travel between them. The first phase of the HS2 megaproject, expected to cost £30 billion,

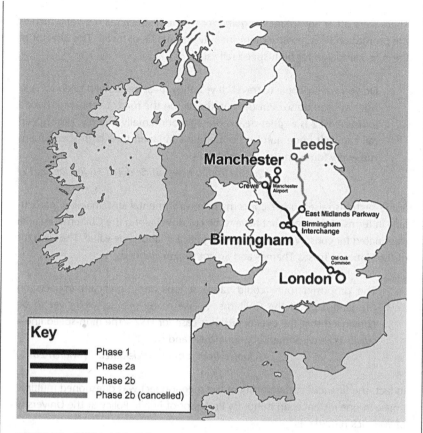

FIGURE 5.3 HS2 initial project map

is proposed to connect London and Birmingham with a 140-mile high-speed rail line. The travel time would be 45 minutes with trains traveling at speeds of 225 mph. The phase 1 of the HS2 megaproject, currently undergoing enabling works, is scheduled to be operational in 2026.

The main objective of the megaproject was to provide 'fast' transport between these two major cities and thereby increase the economic activity in the area. While there was an existing rail connection between the two cities, this megaproject differed in being a 'high-speed' connection. In a statement to the House of Commons, the then UK secretary of state for transport[2] said that:

> A modern and reliable and fast service between our major cities and international gateways befitting the 21st Century will transform the way we travel and promote Britain's economic and social prosperity.
>
> *(Quoted from a news article dated 10 January 2012)*

There were other advantages of going for the high-speed rail network, such as the promotion of an environment friendly alternative of travel. The director of the 'Yes' campaign for high-speed rail said:

> Do you want people to travel? If yes, they must be allowed to do so. And how will you do so? Put them in the air, on the road? Compared to road journeys HS2 is a great deal more environmentally friendly. High speed rail can shift huge numbers of people – in Japan their high-speed trains run every four minutes at peak hours.
>
> *(Quoted from a news article dated 10 January 2012)*

Some sections raised other concerns of environmental sustainability, particularly in terms of the trees that have to be cut down along the Chilterns, an area designated for conservation due to its significant landscape value. The Director of the National Trust's Thames and Solent region claimed:

> The proposed route could cause serious and significant impacts on the landscape of the Chilterns ... Like many people, we're yet to be convinced that the overall business case for HS2 – the high-speed line – stacks up environmentally, financially and socially.
>
> *(Quoted from a news article dated 11 March 2011)*

In fact, the financial sustainability of the megaproject was questioned multiple times. In one instance, an honorary Professor of Public Policy at the University of Warwick remarked:

> This rail link and the 250 mph trains are economically unnecessary and environmentally destructive ... The cost is enormous at a time when public finances are under severe strain, and the business plan is based on over-optimistic forecasts of passengers ... The project does nothing to tackle the immediate problem of overcrowding on trains because it will not be completed for another 15 years.
>
> *(Quoted from a news article dated 14 November 2010)*

The business case for HS2 was even trolled with a spoof set to a famous scene from *Downfall*, a film that charts Adolf Hitler's final days in his Berlin bunker. A news article reported on the video uploaded on YouTube, that was widely shared as below:

> The HS2 parody, which was put online this week, starts with Hitler – in the role of Secretary of State for Transport – saying to his generals: "Don't worry, we have the business case" ... However, one of his staff

replies: "It barely breaks even, despite the insane increase in traffic we forecast." At this, Hitler flies into a rage: "Even with the million jobs we made up you still couldn't get it right. We said it would reduce flights from Heathrow – it won't. It'll only create 10,000 jobs in the Midlands – less than last month's increase in unemployment in Birmingham alone".

(Quoted from a news article dated 4 October 2012)

Reacting to the multiple oppositions to the megaproject, the secretary of state for transport claimed that those who live near the route of HS2 were the ones opposing the megaproject as highlighted below:

I'm afraid none of those objections are valid, though of course I completely understand why those people who live on the line of the route are objecting ... It always happens when you have infrastructure projects, that those who live near where they're being proposed object vigorously and, of course, what they do is to try and draw in wider arguments.

(Quoted from news article dated 19th February 2011)

Some readers of the news agreed that people living near the megaproject are objecting and that the megaproject needs to be built irrespective of this, as seen from one of the comments:

The sooner we build it, the better. There's always negativity around expanding motorways and railways, with the inevitable people in the area saying 'Not in my back yard', but when it's up and running, it becomes the lifeblood of our nation. 40 years ago, there was outrage at the motorway, 100 years ago, it was the railways. Imagine us with neither today!

(Quoted from the comments of a news article dated 28 February 2011)

It is true that people living near the megaproject were unhappy with the proposals. A personal account of a farmer living along the proposed route expressed:

I'm just gutted, and it will be horrifying if it happens. It would ruin the farm and our land won't be worth anything. We don't need High Speed rail and we can't even catch it here anyway.

(Quoted from a news article dated 15 March 2010)

Others joined together with the megaproject affected person as seen from one of the comments on a similar personal account:

And there was me thinking we lived in a democracy! What is the point in working yourself silly to build up a business when the government can come along and make a compulsory purchase of your property if you just happen to be in their way! It's a disgrace. Come on you people of Buckinghamshire, let's fight to keep our countryside!

(Quoted from the comments of a news article dated 14 November 2010)

To hear the concerns of the people, the megaproject conducted a consultation process. It was called as the largest consultation ever undertaken by the spokesperson of the Department for Transport:

This was one of the largest consultations ever undertaken by a government with over 30 events along the line of route attended by tens of thousands of people.

(Quoted from a news article dated 13 November 2011)

The campaigners were not happy with the consultation process undertaken by the Government. A news article reported on the comments by a spokesperson from the Campaign to Protect Rural England:

***[name of person], from the Campaign to Protect Rural England, described the consultation process as "a complete train wreck." He said the consultation amounted to "a single route option, which the government has already made up its mind to favour" and the country needed a "fair, open and informed debate about HSR [High Speed Rail 2]".

(Quoted from a news article dated 28 February 2011)

Groups protesting against the megaproject such as AGHAST (Action Groups against High Speed Rail) emerged. Opponents of the megaproject campaigned relentlessly:

Adding carriages to trains and lengthening platforms would ease overcrowding, and upgrading existing trains and tracks would allow trains to run at speeds up to 180mph, Trains at this speed could also run along new tracks which could be built along existing railways or motorways and minimise damage to the environment.

(Quoted from a news article dated 14 November 2010)

However, those in favour of the megaproject highlighted:

When the TGV was going south in France, there was bitter resistance. Parts, like in Britain, were beautiful and protected and lots of people lived alongside. But they did it, and they compensated people properly – which

I think is crucial – and they consulted and in the end they got the lines through. It's not easy, but the idea of not doing it is utter madness. Do we want to live in the 19th century?

(Quoted from a news article dated 10 January 2012)

The latest news is that the government is set to scrap the Eastern leg of HS2 between the Midlands and Leeds. Sources said the impact of scrapping the Leeds leg of HS2 would make journeys longer by 20 minutes. But the government is set to argue the new plans will deliver comparable benefits more quickly and cheaply. One source told BBC political correspondent Nick Eardley they would show an "enormous amount of common sense".

Conservative MPs expressed concerns about the cost of the eastern leg connecting the West Midlands and Leeds, and there were rumours it would be scrapped. Transport Secretary Grant Shapps announced two shorter high-speed routes created in part by upgrading existing lines. One will run between Leeds and Sheffield, another from Birmingham to East Midlands Parkway. The government is also expected to put money aside to explore setting up a tram service for Leeds.

The Northern Powerhouse Partnership, a group of northern local authorities and business leaders, said the decision to scrap the Leeds leg of HS2 was a mistake.

The Partnership director Henri Murison said: "The reported loss of any of the new line on the eastern leg of HS2 is damaging, reducing the benefits of the section being built now between Birmingham and London."

Without the benefits to areas such as Yorkshire and the North East, HS2's status as a project to drive the whole of the UK is undermined considerably.

A proposed Northern Powerhouse route from Leeds to Manchester is now expected to be made up of some new line, but it will mostly consist of upgrades to the existing track, as shown in Figure 5.4.

The new track on the route will not allow high-speed rail travel. The route is not expected to go via Bradford, a key request of many in the city and surrounding area. Naz Shah, the Labour MP for Bradford West, tweeted:

This is Boris [Johnson] pulling the whole damn rug from under our feet and ripping up the floor behind him!

The government's view is that the alternative option for HS2 would still make journey times faster, but deliver benefits more quickly. There's clearly a cost element too. But it doesn't bring the same benefits when it comes to capacity.

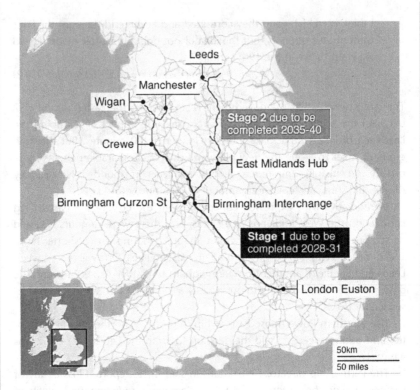

FIGURE 5.4 Updated map of HS2

Meanwhile, there's disappointment in cities such as Bradford. They see the delivery of Northern Powerhouse Rail, including a new line from Leeds to Manchester via Bradford, as crucial to regeneration and creating opportunities for their residents.

The shadow transport secretary, Jim McMahon, accused the government of trying to back out of promises made on badly-needed major infrastructure projects, and described the reported plans as "half-baked and repackaged". Conservative MP Kevin Hollinrake, who represents Thirsk and Malton in North Yorkshire, told one newspaper that the down-sized plans indicated the government was "not willing to put our money where our mouth is".

Exercise

From the above case study, please answer the following questions:

1. What are the different HS2 megaproject narratives observed?
2. What does the narrative imply for the megaproject, the people involved, and the processes followed?

3. Identify the promoters and protesters of each project narrative.
4. How are narratives being maintained and how are they being challenged?
5. List some other arenas (other than news articles) where project narratives can be observed.

Sources

BBC. (2010). High-speed rail plans announced by government. http://news.bbc.co.uk/2/hi/uk_news/8561286.stm (Accessed on 31 July 2022).

BBC. (2011a). High-speed rail campaigners gather for national meeting. https://www.bbc.co.uk/news/uk-england-12514335 (Accessed on 31 July 2022).

BBC. (2011b). Government starts high-speed rail consultation. https://www.bbc.co.uk/news/uk-politics-12591464 (Accessed on 31 July 2022).

BBC. (2012). HS2: High-speed rail go-ahead prompts mixed reaction. https://www.bbc.com/news/uk-16494354 (Accessed on 31 July 2022).

Bracchi, P. (2011). Fury for home owners booted out to make room for a high-speed rail link…but minister behind it halted a similar project in his own back yard. https://www.dailymail.co.uk/news/article-2060575/Home-owners-booted-make-room-rail-link-Buckinghamshire.html (Accessed on 31 July 2022).

Dunhill, L. (2010a). Woman stunned by plans to bulldoze her Frith Hill home for High Speed trains. https://www.bucksfreepress.co.uk/news/5061521.woman-stunned-by-plans-to-bulldoze-her-frith-hill-home-for-high-speed-trains/ (Accessed on 31 July 2022).

Dunhill, L. (2010b). Woman stunned by plans to bulldoze her Frith Hill home for High Speed trains. https://www.bucksfreepress.co.uk/news/5061521.woman-stunned-by-plans-to-bulldoze-her-frith-hill-home-for-high-speed-trains/ (Accessed on 31 July 2022).

Hall, J. (2012). HS2 rail link gets 'Hitler parody'. https://www.telegraph.co.uk/news/uknews/9586612/HS2-rail-link-gets-Hitler-parody.html (Accessed on 31 July 2022).

Harrison, D. (2010). Middle England on the march as revolt over 250mph rail link grows. https://www.telegraph.co.uk/news/uknews/road-and-rail-transport/8131013/Middle-England-on-the-march-as-revolt-over-250mph-rail-link-grows.html (Accessed on 31 July 2022).

Rigby, J. (2012). HS2: What's in it for you?. https://www.channel4.com/news/hs2-whats-in-it-for-you (Accessed on 31 July 2022).

Notes

1 The video titled 'Downfall of HS2' was found through a YouTube search after its mention in the news article. It is available at https://www.youtube.com/watch?v=7WUuagYAj_w and was accessed on 20 July 2022.
2 The House of Commons is the lower house of the Parliament of the United Kingdom. Like the upper house, the House of Lords, it meets in the Palace of Westminster.

References

Andrews, M. (2002). Introduction: Counter-narratives and the power to oppose. *Narrative Inquiry*, 12(1), 1–6.

Boje, D. M. (2008). *Storytelling organizations*. London: Sage.

Dailey, S. L., & Browning, L. (2014). Retelling stories in organizations: Understanding the functions of narrative repetition. *Academy of Management Review*, 39(1), 22–43.

Denning, S. (2005). Transformational innovation: A journey by narrative. *Strategy & Leadership*, 33(3), 11–16.

Di Maddaloni, F., & Davis, K. (2017). The influence of local community stakeholders in megaprojects: Rethinking their inclusiveness to improve project performance. *International Journal of Project Management*, 35(8), 1537–1556.

Flyvbjerg, B. (1998). *Rationality and power: Democracy in practice*. Chicago, IL: University of Chicago Press.

Ganz, M. (2011). Public narrative, collective action, and power. In S. Odugbemi & T. Lee (eds.), *Accountability through public opinion: From inertia to public action* (pp. 273–289). Washington: The World Bank.

Garud, R., & Turunen, M. (2017). The banality of organizational innovations: Embracing the substance-process duality. *Innovation*, 19(1), 31–38.

Granqvist, N., Grodal, S., & Woolley, J. L. (2013). Hedging your bets: Explaining executives' market labeling strategies in nanotechnology. *Organization Science*, 24(2), 395–413.

Grayson, K. (1997). Special session summary narrative theory and consumer research: Theoretical and methodological perspectives In M. Brucks & D. J. MacInnis (eds.), *NA – Advances in consumer research* (Vol. 24, pp. 67–70). Provo, UT: Association for Consumer Research.

Jarzabkowski, P. A., & Lê, J. K. (2017). We have to do this and that? You must be joking: Constructing and responding to paradox through humor. *Organization Studies*, 38(3–4), 433–462.

Jaspercopping (2010). High-speed line noise will affect 50,000 people. *The Telegraph*. https://www.telegraph.co.uk/news/uknews/road-and-rail-transport/8212095/High-speed-line-noise-will-affect-50000-people.html (accessed on 31 July 2020).

Kotter, J. P. (2012). *Leading change*. Boston, MA: Harvard Business Press.

Lee, D., Hosanagar, K., & Nair, H. S. (2018). Advertising content and consumer engagement on social media: Evidence from Facebook. *Management Science*, 64(11), 5105–5131.

Lim, K. H., Sia, C. L., Lee, M. K., & Benbasat, I. (2006). Do I trust you online, and if so, will I buy? An empirical study of two trust-building strategies. *Journal of Management Information Systems*, 23(2), 233–266.

Love, P. E., & Ahiaga-Dagbui, D. D. (2018). Debunking fake news in a post-truth era: The plausible untruths of cost underestimation in transport infrastructure projects. *Transportation Research Part A: Policy and Practice*, 113, 357–368.

Mathur, S., Ninan, J., Vuorinen, L., Ke, Y., & Sankaran, S. (2021). An exploratory study of the use of social media to assess benefits realization in transport infrastructure projects. *Project Leadership and Society*, 2, 1–10.

Ninan, J., Clegg, S., & Mahalingam, A. (2019). Branding and governmentality for infrastructure megaprojects: The role of social media. *International Journal of Project Management*, 37(1), 59–72.

Ninan, J., Mahalingam, A., & Clegg, S., (2022). Power in news media: Framing strategies and effects in infrastructure projects. *International Journal of Project Management*, 40(1), 28–39.

Ninan, J., & Sergeeva, N. (2021). Labyrinth of labels: Narrative constructions of promoters and protesters in megaprojects. *International Journal of Project Management*, 39(5), 496–506.

Olander, S., & Landin, A. (2008). A comparative study of factors affecting the external stakeholder management process. *Construction Management and Economics*, 26(6), 553–561.

Pontikes, E. G. (2018). Category strategy for firm advantage. *Strategy Science*, 3(4), 620–631.

Ponton, H., Osborne, A., Thompson, N., & Greenwood, D. (2020). The power of humour to unite and divide: A case study of design coordination meetings in construction. *Construction Management and Economics*, 38(1), 32–54.

Riessman, C. K. (2002). Narrative analysis. In A. M. Huberman & M. B. Miles (eds.), *The qualitative researcher's companion* (pp. 17–70). Thousand Oaks, CA: Sage.

Sergeeva, N. (2019). *Making sense of innovation in the built environment*. Oxfordshire, UK: Routledge.

Sergeeva, N., & Green, S. D. (2019). Managerial identity work in action: Formalised narratives and anecdotal stories of innovation. *Construction Management and Economics*, 37(10), 604–623.

Sorsa, V., & Vaara, E. (2020). How can pluralistic organizations proceed with strategic change? A processual account of rhetorical contestation, convergence, and partial agreement in a nordic city organization. *Organization Science*, 31(4), 839–864.

Sturup, S. (2009). Mega projects and governmentality. *World Academy of Science, Engineering and Technology*, 3(6), 892–901.

Turner, J. R., Lecoeuvre, L., Sankaran, S., & Er, M. (2019). Marketing for the project: Project marketing by the contractor. *International Journal of Managing Projects in Business*, 12(1), 211–227.

Vaara, E., Sonenshein, S., & Boje, D. (2016). Narratives as sources of stability and change in organizations: Approaches and directions for future research. *Academy of Management Annals*, 10(1), 495–560.

Van Marrewijk, A., Clegg, S. R., Pitsis, T. S., & Veenswijk, M. (2008). Managing public–private megaprojects: Paradoxes, complexity, and project design. *International Journal of Project Management*, 26(6), 591–600.

Weick, K. E. (1988). Enacted sensemaking in crisis situations. *Journal of Management Studies*, 25(4), 305–317.

Zakhem, A. (2007). Stakeholder management capability: A discourse-theoretical approach. *Journal of Business Ethics*, 79(4), 395–405.

6

NARRATING AND STORYTELLING

6.1 Introduction

In this chapter we discuss the processes of narrating and storytelling, their inter-actions and the implications of these on practices of megaprojects. Managers are expected to narrate and tell stories about themselves and others, their experiences and work practices, situations and events. This is evident in spoken, textual and visual forms of narrating and storytelling which are pivotal processes that shape and promote megaprojects.

6.2 Interaction between narrating and storytelling

In the first chapter of the book, we outlined the distinctive characteristics of nar-ratives and differentiate them from stories. Whilst narratives tend to be scripted, performative and strategic in nature, stories are more entertaining, personalised and emotional in nature (Sergeeva & Green, 2019). In a similar vein, we differentiate between narrating and storytelling, arguing that there is a need for both in organ-isational life. There is an ongoing process of interaction between narrating and storytelling. People make sense of their lives and experiences and communicate via narratives and stories. Managers and leaders are expected to use both narrating and storytelling and to oscillate between the two in conversations, when give speeches, presentations and in videos.

Narrating is a process of articulating and using narratives that embodies a degree of coherence and unity of purpose together with connotations of performative intent (Vaara et al., 2016). It is associated with a process of narrative repetition in organisations (Dailey & Browning, 2014). Narrating hence may carry impor-tant messages at the level of the firm with reference to national and international

DOI: 10.1201/9781003248378-6

agendas (Sergeeva & Winch, 2020). Managers narrate their messages, since such people bear responsibility for formulating and disseminating organisational strategies and future vision (Sims, 2003; Sonenshein, 2010). For example, Abolafia (2010) demonstrates the ways in which elite policy makers use plotted, plausible and repeated narratives to shape the reactions of those in their environment and to develop strategies. Rehearsed, often-dominant narratives also invariably play an important role in legitimising the advocated actions. Organisations need a coherent and consistent narrating process because narratives represent attempts to impose order, help to overcome resistance to change and achieve common understanding and vision. In contrast to narrating, storytelling is understood as a process of sharing stories about personal experiences, themselves and others, and events told for entertainment purposes.

Storytelling has a long history in organisation and management studies (Boje, 2008, 2011; Gabriel, 1995, 2000). Stories are the means by which people ascribe meaning to their experiences, in which these are understood as beliefs, meanings, interpretations and actions. It is through stories of everyday life experience that we can learn how people feel and make sense of themselves, others, their work practices, and situations. Storytelling is a powerful lens through which we can explore perceptions of people and what they actually do, and therefore the process of sharing stories is an important part of organisational life as it enables people to learn from each other. People are naturally good at sharing stories, and it often helps them to reflect on the past, the present, and think about the future and other related issues that they may not have considered before. As discussed in Chapter 4, stories are personalised and usually emotional in nature. As a listener, it is important to put care, respect and empathy into understanding speakers. It is an emotional intelligence competence to be able to understand speakers, their emotions through stories and storytelling. For example, in an interview or a conversation situation asking relevant questions which make people feel comfortable and open to talk and share stories is very important.

A process of oscillating between narrating and storytelling reflects continuous processes of interactions in organisations between formal and informal roles, stability and repetition on the one hand and change and transformation on the other hand. A process of interaction between narrating and storytelling is a source of both stability and change in organisations. Stories can be employed by both the promoters and the protesters of a megaproject (Ninan & Sergeeva, 2022).

6.3 Formal and informal roles and identities

Whilst narrating process is associated with more formal roles in organisations, storytelling is more often associated with more informal roles and self-identities. Managers are expected to give a coherent narrative of organisational performance for their staff. But they also continuously and spontaneously construct stories of what is happening in their lives, as well as revising them and imagining the future.

Stories and storytelling, and a sense of humour, are important means through which individuals strive to make sense of narratives and narration. They hence comprise "fragments" of organisational discourse that craft identities and interests in time and space. In organising, individuals are using both storytelling and narrating.

The role stories and storytelling play in the social construction of individual and work-related identities is widely recognised (Alvesson, 2010; Bruner, 1991; Thomas & Davies, 2005). Individual identity or self-identity is a sense of self: Who am I? What do I do? An individual's work identity refers to work-based identity consisting of a combination of organisational, occupational and personal identities that shape the informal roles that an individual adopts, and the ways they behave when performing their work (Ninan & Sergeeva, 2021).

Vaara and Tienari (2011: 370) define stories as "fragments of organizational discourse that construct identities and interests in time and space". By telling stories or self-narratives individuals seek to bolster their identities, in the eyes of both themselves and others (Alvesson et al., 2008; Baumeister & Newman, 1994; Brown, 2015). Ibarra and Barbulescu (2010) and Järventie-Thesleff and Tienari (2016) focus on the way people in organisations engage in transitions within and between informal roles, and the implications of these transitions for their self-identities. Sergeeva and Green (2019) conducted narrative interviews with senior executives of construction firms in the UK and show the ways they oscillate between performative narratives of innovation and more personalised stories derived from their own experience. They used a metaphor of 'on and off soapbox' to demonstrate a process of oscillating between narratives and stories by senior executives. They also show the ways senior executives self-ascribe themselves with informal roles such as 'honest John', 'maverick', 'action man', 'portrayer of pride', 'lone enlightened thinker', 'organisational scriptwriter' and 'empathetic reflective practitioner'. In a later work, Sergeeva and Kortantamer (2021) conducted life-story interviews with Chief Executive Officers (CEOs) of public project-based organisations in the UK construction industry and reveal a dialogical process between more coherent narratives and personalised stories articulated by the leaders. They discuss continued interweaving stories of self and the context, and ongoing efforts to cope with relational anxieties, authenticity and self-identity struggle. Sergeeva (2022) studied storytelling of sustainability, showing that when talking about sustainability practices and activities, practitioners presented themselves with work-related identities. They emphasised their motivation, passion and enthusiasm for sustainability in addressing the macro-level narratives of sustainability. Among the work-related identities discussed are activists, supporters and facilitators of sustainability. Activists are individuals who are proactively driving the sustainability agenda among colleagues. They are champions of change. Supporters are individuals who support the sustainability agenda and provide guidance to other stakeholders and employees for implementing sustainability practices. Finally, facilitators of sustainability are individuals who guide and stimulate the team to work towards the achievement of sustainability goals. Facilitators play an important job in helping groups of people to make decisions. They provoke the group to stimulate discussions; they are key people in meetings and events. Anyone in an organisation can play informal roles and have work-related identities.

6.4 Narrating and storytelling in megaprojects

There is a growing interest in studying narratives and stories, and the activities of both narrating and storytelling, within the context of megaprojects. For example, Boddy and Paton (2004) analyse accounts of three major projects and demonstrate how competing narratives are managed, which reflect the diverse realities within which most projects take place. Manning and Bejarano (2017) explore the ways project histories and potential futures are framed and interlinked in narratives to appeal to project sponsors. They found that the imagined futures of a project were narrated in different styles, one style focuses on immediate future steps whereas the other highlights the long-term vision, but they refer to storytelling/stories and narrating/narratives interchangeably without making any distinction between them. Havermans et al. (2015) build upon narrative theory and studied individual stories mobilised by project leaders about their experiences and events, arguing that language is constitutive of organisational reality rather than merely representative. Enninga and van der Lugt (2016) further investigate the role narratives play in leading innovation projects and the ways in which innovation project leaders use stories in practice, again referring to narratives and stories interchangeably. In our work we distinguish between the two and argue that narrating tends to be deliberately used in project organising for different purposes, including to create certain futures, persuade investors, legitimise actions, and promote projects both internally and externally (Sergeeva & Winch, 2021). Sergeeva and Davies (2021) show that it is the most personalised and reflective stories that are most effective when used by the megaproject leader. They present a life story of the CEO of the HS2 megaproject who sees himself as authentic leader. In his story they found a strong connection between self-identity and organisational identity.

Narrating and storytelling are crucial processes that shape, affect and promote megaprojects. These processes are found to be important throughout a project's lifecycle. As discussed in Chapter 2, at the shaping stage of a project there is an ongoing process of narrating; a range of different project proposals may be discussed before one is selected, approved and justified. The process of narrating continues as there is a need to sustain project narrative among project stakeholders. Project leaders and their teams are exercising a process of narrating by mobilising and communicating project narratives at different occasions and situations. Storytelling among the members of project teams, and others who are affected (such as external stakeholders share stories about their experiences with a project. Stories get heart and become circulated on social media, websites, newspapers and other sources. There is an ongoing process of storytelling in the megaproject process. Project leaders and team members narrate and tell stories which make the project experience interesting, memorable and engaging. During the construction phase, project managers narrate and tell stories about work practices to improve project management performance. For example, Datta et al. (2020) record how stories about project process with visualisation can improve the level of comprehension among non-technical senior leaders and bring stakeholders together. During the post-project evaluation phase, stories of project experiences, successes and learning from failures become promoted more widely. Narrative and storytelling are part of a legacy of megaprojects and the whole ecosystem of projects. These processes create and shape the future, and impact both internal and external stakeholders (Ninan et al., 2021).

6.5 Conclusion

In this chapter we discuss the importance of narrating and storytelling in organisational life, and specifically in the context of megaprojects. There is an ongoing process of interaction between narrating and storytelling, and such interaction plays an important role in projects and people's experiences. We discuss these processes in relation to formal and informal roles in organisations, providing some examples. We highlight that both narrating and storytelling are needed throughout a project's lifecycle. Project leaders and team members narrate and tell stories which makes the project experience interesting, memorable and engaging.

CASE STUDY 6: CRIMEAN BRIDGE, RUSSIAN FEDERATION

History of the Crimean bridge

It had been recognised for a long time that there was a need to connect Kerch and Taman with a ferry. Projects were constantly proposed, but all were delayed for one reason or another. Either there were insufficient funds, or a the proposal was turned down. There was even a proposal to build a tunnel under the strait, but nothing came of it. Various options for the route were proposed. However, many of these could not be implemented due to the complexities of the local coastline. After one of the project proposals was accepted, construction work began.

The Crimean bridge has opened officially on 15 May 2018. It connects the Crimean peninsula with the mainland. It is located on the side of the Republic of Crimea – near the city of Kerch. The Crimean (or Kerch) bridge is both a highway and a railway bridge over the Kerch strait that connecting two peninsulas, Taman and Crimea, from the Krasnodar Territory – in the area of the village of Taman, Temryuk District. The route of the transport crossing passes in the alignment of the island of Tuzla and the Tuzla Spit. The Crimean bridge, like any other major crossing, is considered a strategic facility. Figure 6.1 shows the completed Crimean bridge project.

FIGURE 6.1   Crimean bridge

On the anniversary of the opening of automobile traffic on the Kerch bridge, the Head of Crimea, Sergey Aksenov, spoke about its benefits for the peninsula:

> The most obvious result is a sharp, almost 30 per cent, growth in tourist traffic following the results of last year. Over the past holidays, it has grown by 40 per cent compared to the level of 2018. The bridge itself has become one of the main attractions, an important tourist attraction, a "visiting card" of the region. After the bridge was put into operation, the average monthly intensity of cargo transportation between the mainland and the peninsula increased by almost 2.5 times. The number of cars that have passed in both directions has already reached five million. Moreover, the traffic flow is equally intense in both directions. That is, the bridge is in demand on both sides of the Kerch strait. The construction of the bridge gave a powerful impetus to the development of the entire transport infrastructure of the peninsula, and this is one of the foundations of the new economy of the peninsula … The opening of railway traffic on the bridge at the end of this year will allow most of the products to be delivered by rail, which will affect its final cost.

Crimean political scientist, Yevgenia Goryunova, believes that the bridge across the Kerch strait did not have such a positive impact on the standard of living in

Crimea, in contrast to what the local authorities had previously promised the Crimeans:

> This is unlikely to have changed the living conditions of the Crimeans because what was expected from the bridge – price reduction – did not happen. Moreover, today they say that the opening of the railway part will not reduce prices. A little more tourists to the territory of Crimea really began to travel across the bridge, so far it has not given any more positive results.

Experts claim the bridge is very expensive when compared with other similar infrastructure projects. According to Yevgenia Goryunova:

> Contractors preferred to bring people from Siberia, from other regions of Russia on a rotational basis – sometimes they get paid, sometimes not. They saved on labour by not hiring Crimeans. But, firstly, the bridge has a symbolic meaning, and secondly, it is needed to transfer military equipment. No one thought much about prices – everyone knew perfectly well that they would not decrease, that neither road nor rail transport would help there until large Russian trading companies entered Crimea. They do not want to enter Crimea because they are afraid of sanctions.

Yuri Medovar, an independent expert, and a student of geological and mineralogical sciences, is convinced that a lot more money has been spent from the budget to maintain the Kerch bridge:

> The whole bridge is one big problem. I believe that the operating costs of maintaining this bridge will soon outweigh its cost. There will be a "cut" of the budget into the pockets of those affiliated companies. The amount spent on the Kerch bridge could be used as a free crossing across the strait for 50 years – both trains and cars could cross on ferries. Now the bridge is sinking somewhere, rising somewhere, it will be uninterrupted. This structure, from the point of view of the geological structure of the strait, should never be there. This is a violation of all building norms of the Soviet Union. Near the bridge there is a mud volcano that erupts about once every 8–10 years. If this happens, there will be nothing left of the bridge.

According to Yuri Medovar, the launch of the railway part of the bridge and the beginning of train services will be a serious test for the entire structure:

> The road bridge is a light structure. At the railway bridge, a large arch, about 10,000 tons, stands on two supports, and imagine that a train of 60 wagons will go – that's another 6,000 tons. 10 KAMAZ vehicles passed – without cargo, light, they are 10–15 tons each, the total load

FIGURE 6.2 Structure of the Crimean bridge

is 150 tons. Bridge acceptance is when there is a concentrated load on the entire length of the bridge. Bridges were accepted in Prague – they put tanks one to one, looked at the draft. Here is the acceptance of the bridge into operation.

Figure 6.2 shows the structure of the Crimean bridge.

Narrating environmental sustainability

It is argued that the construction of the Crimean bridge did not affect the ecology of the bay. During the construction of the bridge, reliable protection of the region's flora and fauna was guaranteed. This was told by Murad Kerimov, the Deputy Minister of Natural Resources and Ecology:

> Back in January 2015, a working expert group was created under the Ministry of Natural Resources of Russia, which included specialists from the Russian Academy of Sciences, Moscow State University, research institutes and public environmental organisations, specially protected natural areas. Several sections worked in the expert group: protected areas, protection of plants and animals, marine protection, waste management, hydrometeorology and environmental monitoring.

In the summer of 2015, the first public hearings on the program of environmental surveys were held, their examination was carried out, and then the engineering and field studies themselves, Red Data Book animals and plants were relocated to the territory of the reserve. In November, they received the conclusion of the environmental impact assessment of the project and began the first work:

> Two species of cetaceans live in the Kerch Strait – bottlenose dolphin and Azov porpoise. They use the Kerch Strait in different ways. The bottlenose dolphin lives in it in the summer; the Azovka migrates between Azov and the Black Sea, where it goes for the winter. There is also a white-barreled dolphin, or an ordinary one, but it swims into the bay extremely rarely. In the first months of preparation for construction, when sappers were working in the water area and accidental detonation of explosive objects from the times of the Great Patriotic War was possible, acoustic scaring was used. Azovka do not like noise and never accompany ships. They could experience some discomfort during the period of mass pile driving, nevertheless, in spring and autumn they continued to successfully migrate through the strait.
>
> *(Murad Kerimov)*

According to experts, bottlenose dolphins turned out to be the most stress-resistant: even the noisy construction phase had almost no effect on them. It is expected that in the future, dolphins will inhabit thousands of piles in the strait – a kind of artificial reef, which will increase the biological productivity of the water area.

> There is evidence of an increase in the population of the Azov herd of porpoises. Compared with 2001–2002, their number has more than tripled and now exceeds 17,000 individuals.
>
> *(Murad Kerimov)*

> In 2017, on behalf of the Ministry of Natural Resources, an expedition of the All-Russian Research Institute "Ecology" was carried out with the participation of employees of the Crimean branch and invited experts. A series of ship surveys was carried out, the number of migrating dolphins was determined, coastal observations and surveys of the local population and fishermen were conducted along the way. Work continued in 2018. A census of dolphins in the Russian waters of the Black Sea was planned, which has not been fully carried out for three decades. Scientists have confirmed that the construction of the bridge does not have a negative impact on the population of dolphins using the Kerch Strait. In the winter of 2017–2018, there was a delay in the migration of fish, dolphins and birds, but this was caused by abnormally warm weather and was not related to construction.
>
> *(Murad Kerimov)*

Migratory routes of birds, among which there are rare species, turned out to be in the construction zone. Among the main risks, experts called the impact of noise on their habitats, they feared the displacement of birds to other territories and even the death of other rare species. To avoid this, noise barriers, feeders, artificial nests on the water and on land were installed. Translocation of three red book species of reptiles was carried out. Ecologists also identified 8 rare and protected plant species, all of which were transplanted from the construction zone to compensation plots.

(Murad Kerimov)

When developing the draft environmental impact assessment (EIA) in 2015, territorial zoning was carried out, which made it possible to determine the boundaries of the construction of the Crimean bridge. Part of the temporary infrastructure facilities was located within the boundaries of the Zaporozhye-Tamansky reserve – a specially protected natural area of regional subordination.

(Murad Kerimov)

Summarising the monitoring materials for 2017, it is stated that despite local pollution of the water area, the water quality in the Kerch Strait, as well as in the Taman and Dinskoy bays, meets the standards. Changes in the habitat of bioresources are practically insignificant. According to the results of environmental monitoring for the 4th quarter of 2017, an increase in the total number of birds was noted in the region of the construction of the Crimean bridge.

How the Crimean bridge changed the lives of people in the South of Russia

The news of the Krasnodar region (yuga.ru) decided not only to recall the history, but also to ask about the changes in the residents of Taman and Crimea themselves – entrepreneurs, farmers and ordinary people. Below are a number of their stories.

Irina Gridin, a resident of the village of Ilyich in the Temryuk district, recalls how she and her grandmother waited for many years for the bridge:

Before, the wind would blow more than 20 meters – everything got up. So we went, for example, to Kerch by ferry, but we can't go back, we sat for days, waiting for the storm to end. And now the bridge is accessible in any weather; and Kerch, I think, should only rejoice. This is also my favorite city, I spent all my childhood there. Mom is at work, and once we take the ferry – and to the other side. Do you know how much we loved the ferry? They took a loaf of bread with them – the seagulls were already accustomed, they grabbed bread on the fly. Although the ferry is, of course, romantic, but the bridge is needed. This is a charm, you are no longer afraid that the wind will rise and you will not be able to get home.

What has been going on with us in recent years, what kind of traffic jams were before the crossing; the queue from the crossing was to Zaporizhzhya about 20 km.

When they started talking about the bridge again a few years ago, I didn't believe it anymore. And then, when it became clear that this was a reality, of course, we thought that it would be nice if the bridge to the Crimea went through Chushka. This is work, and how much they would have built. There were several projects, but Chushka was rejected because it was a very narrow spit.

Boris Pavlovsky, CEO of the Crimean company Best-Beton, recalls:

The most difficult thing was at the beginning – no one understood, even among our senior employees, when they were told the cost of concrete on the mainland and the cost of concrete in the Crimea, which varied greatly. We were told that this could not be, they accused us of trying to make money on it. And why are we here? Guys, your material costs 250 rubles, and we have 2000 rubles. In 2014, the peninsula was cut off from everything. Local materials are suitable for civil construction, but they are not suitable for the construction of roads, bridges, railway tunnels and other infrastructure projects.

When they began to build the bridge, many did not understand that there would be problems. Well, it would seem that there is such a thing, sand or gravel cost 250 rubles per ton. But it was impossible to get to us, except through the ferry crossing. Part of the materials was delivered from Rostov by barges, and most of the sand and gravel were transported from the Krasnodar Territory by railway cars to Port Caucasus. And there, the goods were reloaded onto cars that were transported on ferries, including ordinary civilian ones. Then the cargo got to our base, and only after that the movement of material began. According to Pavlovsky, participation in the construction was real manna from heaven for his company:

Before that, there was no work at all – we were on the edge of geography. As a result, they were able to take part in such projects. They worked around the clock. Participated in the construction of all railway approaches to the bridge, in the construction of the tunnel, built all bridges and interchanges from Kerch to Feodosia.

After the road connection was opened, the situation with building materials became much easier – the price of the material fell by almost half. According to the builders, the price for the construction of infrastructure facilities has already fallen and a construction boom has occurred in the Crimea. Only in Kerch, several residential complexes are being built now, a house for migrants has been laid, the construction of a stadium, a kindergarten and a water pumping complex is planned:

> For many, the crossing was a serious obstacle, including for builders. After the bridge appeared, traffic immediately arose here. Immediately you could see the influx of tourists. People are coming and this requires infrastructure development.
>
> We have already opened a hotel right on the waterfront. Part of the hotel has already been launched, and tenants have also been allowed to enter commercial areas, and there are plans to open a chic restaurant. Some of the rooms have been opened, and we have 100 per cent occupancy every day, although the room is not cheap. So the bridge allows you to build such things, and we hope that all this will give a result. I am sure that all the rooms of our hotel will be full.
>
> *(Boris Pavlovsky)*

Pavel Martsinovsky, director of the Atelika hotel chain, explains the situation:

> The coronavirus has affected negatively: we have lost more than a month of the season, and nothing can compensate for this. There is an opinion that hoteliers eventually made money on rising prices – and so this is a myth. Grandmothers renting beds can earn on rising prices. Budgets for the next season are drawn up at the end of the previous one, sales start the previous year, and at the time when the covid disaster happened, everything was actually sold at prices that were before.
>
> If there is such a manna from heaven, as in this case closed Turkey and Egypt, then we do not receive anything from this manna. Our prices have not changed at all this year. Basically everything is pre-bought.

After the opening of the bridge in the Crimean direction, there was an increase in the flow of not only tourists, but also freight.

Trucker, Anatoly, has been traveling all over the country, often going abroad for the past 13 years. This time, a cargo of plastic gutters is being transported to the peninsula:

> A friend of mine carries potatoes to the Crimea from Bryansk. Very fast – one foot here, the other there. He at first shouted that he didn't want to

hear anything about the bridge, but now he calmed down, as they built it. Well, it's really convenient.

Even earlier, truckers recognised the convenience of the bridge by local residents – both Crimeans and residents of Taman. According to Alexander Eremenko, a resident of Kerch, trips to the Krasnodar Territory have ceased to be trips that require preparation:

> Well, really, no one pays me to tell you this – awesome bridge. Previously, going to Temryuk to visit relatives was a responsible matter, not to mention a trip to Krasnodar to visit my brother. It was necessary to look at the weather forecast, in winter, again, you don't train much. Yes, and visiting earlier, when there was a storm, they fell for three or four days. Somehow, I had to go around the Rostov region about eight years ago. Indeed, it was convenient. It's a pity that the bridge didn't affect prices as much as we would like – you have cheaper products in the Kuban.

Kuban producers are already learning to make a profit in the new realities. Seyran Gevorgyan, the owner of the "On Cheese" cheese factory, says:

> People come to us from the nearest villages, but a significant number are auto tourists who go on vacation and visit us with pleasure. There are already regular customers who came last year. And now they are going to the sea again and on the way they will definitely stop by. Buyers come from all walks of life. There are a lot of new ones: those who couldn't go anywhere this year came to us and discovered that we have a lot of interesting things.

Seyran has a successful restaurant business in Moscow. A few years ago, having visited Kuban with friends, I decided to buy vineyards in the village of Zaporozhskaya, located in the north-west of the Taman Peninsula, and opened a cheese factory from scratch:

> At first, I thought to open in the Anapa region, but decided to start a business here. Now I understand that it was the right decision. Everyone looked at me in surprise. Well, of course, someone is doing something incomprehensible in an open field. But everything worked out, and there are big plans for development – we will breed goats, build a spa complex, a pizzeria, a recreation area with guest houses.

Today, the Gevorkyan cheese factory produces dozens of types of high-quality and tasty (we tried!) cheeses. Production volume – up to three tons per month.

Products are supplied to cafes in Simferopol, Sochi, and even to one of the establishments in St. Petersburg. This is not the limit – the owner has big plans to expand supplies to the Crimea and the Krasnodar Territory. Of the visible obstacles, the cheesemaker identifies:

> Electricity is very expensive in the Krasnodar Territory. In Moscow, the enterprise is located within the Boulevard Ring, and electricity is cheaper there than here in the village. This greatly affects development. I don't want to raise the price – I want to make a good product that is accessible to people. We have a very small enterprise, but we have to pay at commercial rates. Therefore, in the future we plan to switch to solar energy.

Alexander Erokhin, melon grower and vegetable grower from the village of Strelka, explains that the opening of the Crimean bridge has seriously changed the lives of local farmers:

> We used to have a problem – the Temryuk district was the farthest, in the corner. At the time of the crossing, Crimea did not take anything from us. Tourists could not get there for three days, what kind of agricultural products are there. They traveled from there to Kherson. Now, when wholesalers from the Crimea fly to the Kuban, we are the first with irrigated land, the first with pepper, eggplant, cabbage.

According to Erokhin, it is important not only to know the technology and grow good products, but you need to have the knowledge and flair to understand what and how much to plant in which year:

> In the Crimea, you need pepper of the Belozerka variety, ivory. Green pepper goes to the Kuban – well, people have such tastes. Therefore, you must know what to plant, when and in what quantity, – explains the vegetable grower. – In the Crimea, there is a shortage of water, but those crops that are difficult to grow without water are very popular – cabbage, peppers with eggplant. So, they come for this to the Temryuk region, to the Crimean and Slavic. Well, why go further?

In addition to watermelons, of which there are more than 200 hybrids on his farm, Alexander Erokhin grows bell peppers, tomatoes, eggplants, cabbage and other vegetables. And in the spring, he made a prediction that this year will be a year of twists. People stayed at home and realised that it is impossible to stay at home without supplies. According to Alexander Erokhin, the profitability of agriculture will only increase every year. The main reason is world hunger. The buyer of such a manufacturer will always find "They will come from the Crimea, and from Krasnodar. If it tastes good."

Exercise

1. How was a narrative about project need crafted?
2. What are the project narratives about need and benefits by promoters and what are the counter-narratives advanced by protesters?
3. Identify narratives and stories about the project.
4. How are project narratives connected with people's stories?

Sources

Berezina, E. (2018). The construction of Crimean bridge did not affect the ecology of the bay. *RGRU news.* https://rg.ru/2018/06/03/reg-ufo/stroitelstvo-krymskogo-mosta-ne-povliialo-na-ekologiiu-zaliva.html (Accessed on 19 August 2022).

Dereza, V., & Sineok, E. (2020). "At first he shouted that he didn't want to hear anything about the bridge, but now he calmed down." How the Crimean Bridge changed the lives of people in the south. *Yuga.ru news.* https://www.yuga.ru/articles/society/9188.html (Accessed on 19 August 2022).

Kazarin, P. (2015). The reliability of the bridge across the Kerch strait is doubtful. *Underground Expert news paper.* https://undergroundexpert.info/opyt-podzemnogo-stroitelstva/intervyu-s-ekspertami/most-kerch-nadezhnost/ (Accessed on 19 August 2022).

The Crimean Bridge – The Kerch. (2022). *Russia Travel.* https://russia.travel/objects/330378/ (Accessed on 19 August 2022).

Yankovskiy, A. (2019). "The whole bridge is one continuous problem": A year after launch of the automobile part of the Kerch bridge. *Crimea. Realities.* https://ru.krymr.com/a/godovshina-avtomobilnoy-chasti-kerchenskogo-mosta/29942911.html (Accessed on 19 August 2022).

References

Abolafia, M. Y. (2010). Narrative construction as sensemaking: How a central bank thinks. *Organization Studies,* 31(3), 349–367.

Alvesson, M. (2010). Self-doubters, strugglers, storytellers, suffers and others: Images of self-identities in organization studies. *Human Relations,* 63(2), 193–217.

Alvesson, M., Ashcraft, K. L., & Thomas, R. (2008). Identity matters: Reflection on the construction of identity scholarship in organization studies. *Organization,* 15(1), 5–28.

Baumeister, R. F., & Newman, L. S. (1994). How stories make sense of personal experiences. *Personality and Social Psychology Bulletin,* 20(6), 676–690.

Boddy, D., & Paton, R. (2004). Responding to competing narratives: Lessons for project managers. *International Journal of Project Management,* 22(3), 225–233.

Boje, D. M. (2008). *Storytelling organizations.* London: Sage Publications.

Boje, D. M. (2011). *Storytelling and the future of organizations: An antenarrative handbook.* Abingdon, UK: Taylor & Francis.

Brown, A. D. (2015). Identities and identity work in organizations. *International Journal of Management Reviews,* 17(1), 20–40.

Bruner, J (1991). The narrative construction of reality. *Critical Inquiry,* 18(1), 1–21.

Datta, A., Ninan, J., & Sankaran, S., (2020). 4D visualization to bridge the knowing-doing gap in megaprojects: An Australian case study. *Construction Economics and Building,* 20(4), 25–41.

Dailey, S. L., & Browning, L. (2014). Retelling stories in organizations: Understanding the functions of narrative repetition. *Academy of Management Review*, 39(1), 22–43.

Enninga, T., & van der Lugt, R. (2016). The innovation journey and the skipper of the raft: About the role of narratives in innovation project leadership. *Project Management Journal*, 47(2), 103–114.

Gabriel, Y. (1995). The unmanaged organization: Stories, fantasies and subjectivity. *Organization Studies*, 16(3), 477–501.

Gabriel, Y. (2000). *Storytelling in organizations: Facts, fictions, and fantasies.* Oxford: Oxford University Press.

Havermans, L. A., Keegan, A., & Den Hartog, D. N. (2015). Choosing your words carefully: Leaders' narratives of complex emergent problem resolution. *International Journal of Project Management*, 33(5), 973–984.

Ibarra, H., & Barbulescu R (2010). Identity as narrative: Prevalence, effectiveness, and consequences of narrative identity work in macro work role transition. *Academy of Management Review*, 35(1), 135–154.

Järventie-Thesleff, R., & Tienari, J. (2016). Roles as mediators in identity work. *Organization Studies*, 37(2), 237–265.

Manning, S., & Bejarano, T. A. (2017). Convincing the crowd: Entrepreneurial storytelling in crowdfunding campaigns. *Strategic Organization*, 15(2), 194–219.

Ninan, J., Mahalingam, A., & Clegg, S. (2021). Asset creation team rationalities and strategic discourses: Evidence from India. *Infrastructure Asset Management*, 8(2), 1–10.

Ninan, J., & Sergeeva, N. (2021). Labyrinth of labels: Narrative constructions of promoters and protesters in megaprojects. *International Journal of Project Management*, 39(5), 496–506.

Ninan, J., & Sergeeva, N. (2022). Mobilizing megaproject narratives for external stakeholders: A study of narrative instruments and processes. *Project Management Journal*, 53(5), 520–540.

Sergeeva, N. (2022). *Sustainability: Inclusive storytelling to aid sustainability development goals.* Buckinghamshire, UK: Association for Project Management (APM).

Sergeeva, N., & Davies, A. (2021). Storytelling from the authentic leader of High Speed 2 (HS2) Ltd. Infrastructure megaproject in the United Kingdom. In N. Drouin, S. Sankaran, A. V. Marrewijk, & R. Müller (eds.), *Managing mega-projects: Leadership of major infrastructure projects. New horizons in organization studies* (pp. 47–61). Cheltenham, UK: Edward Elgar.

Sergeeva, N., & Green, S. D. (2019). Managerial identity work in action: Formalised narratives and anecdotal stories of innovation. *Construction Management and Economics*, 37(10), 604–623.

Sergeeva, N., & Kortantamer, D. (2021). Enriching the concept of authentic leadership in project-based organisations through the lens of life-stories and self-identities. *International Journal of Project Management*, 39(7), 815–825.

Sergeeva, N., & Winch, G. M. (2020). Narrative interactions: How project-based firms respond to Government narratives of innovation. *International Journal of Project Management*, 38(6), 379–387.

Sergeeva, N., & Winch, G. M. (2021). Project narratives that potentially perform and change the future. *Project Management Journal*, 52(3), 264–277.

Sims, D. (2003). Between the millstones: A narrative account of the vulnerability of middle managers' storying. *Human Relations*, 56(10), 1195–1211.

Sonenshein, S. (2010). We're changing – Or are we? Untangling the role of progressive, regressive, and stability narratives during strategic change implementation. *Academy of Management Journal*, 53(3), 477–512.

Thomas, R., & Davies, A. (2005). Theorising the micro-politics of resistance: Discourses of change and professional identities in the UK public services. *Organization Studies*, 26(5), 683–706.

Vaara, E., Sonenshein, S., & Boje, D. (2016). Sources of stability and change in organizations: Approaches and directions for future research. *The Academy of Management Annals*, 10(1): 495–560.

Vaara, E., & Tienari, J. (2011). On the narrative construction of multinational corporations: An antenarrative analysis of legitimation and resistance in a cross-border merger. *Organization Science*, 22(2), 370–390.

7

MOTIVATIONS AND SENSE OF PRIDE

7.1 Introduction

Narratives can bring together all of the stakeholders in a project, an important factor in securing successful delivery. While some narratives inhibit mindful action, others can facilitate it (Ganz, 2011), and it is this facilitating power of narrative that we uncover in this chapter. Along with bringing together promoters and protesters, narratives can also influence the employees of the megaproject, helping them create a shared vision for the project. In this chapter, we discuss some of the motivations highlighted by the employees of a megaproject and understand why these motivations exist.

7.2 Motivations in megaprojects

Megaproject team are internal stakeholders who are involved in the creation of the megaproject, including the government, contractors, financers, designers, consultants, etc. The construction of a megaproject is dependent on many factors, such as the expertise of the team, the resources available to them, and their rationalities or motivations and beliefs. Zuashkiani et al. (2014) note the role of project team expertise, such as business, financial, engineering and operational know-how in extracting maximum value from an infrastructure asset. Ninan et al. (2019) highlight how the resources available to the project team such as fund discretion, recruitment discretion and government backing determine project management actions in managing external stakeholders. While the expertise of the project team and the resources available to them are explored in previous works, there has been little discussion of the rationalities or motivations that shape these actions in the literature.

In the past, the decisions of the project team and the rationalities that shape them have been influential in megaproject research. For example, strategic misrepresentation

DOI: 10.1201/9781003248378-7

involves the project team systematically and repeatedly underestimating the cost of the project and overestimating its benefits (Wachs, 1989). These forms of accounting are then strategically used to justify the selection of the megaproject (Flyvbjerg, 2008), leading to many megaprojects being critically compromised by cost and schedule overruns and the 'survival of the unfittest'. Scholars tend to explain strategic misrepresentation based on optimism bias, technological sublime, or corruption in the decision. With optimism bias, the project team behaviourally tends to optimism regarding success, downplaying the risks of failure and neglecting past statistics in evaluating current plans (Kahneman & Lovallo, 1993). The technological sublime refers to engineers and designers' interest in creating and designing megaproject technologies and having their name associated with a major achievement, as discussed earlier. Corruption occurs when illegitimate means are used to further the interests of infrastructure megaprojects (Locatelli et al., 2017). In contrast to explanations such as optimism bias, technological sublime or corruption, we consider the project team's decisions as embedded in repertoires of socially and organisationally available discursive rationalities or narratives. We explore how these rationalities come into being and how they are maintained.

The project team rationalities of pride in the asset, the importance of the asset, the overemphasis on time, and the acceptability of public inconvenience, are discussed below.

7.2.1 Pride in the megaproject

The project team is inherently very proud of being associated with the megaproject and this is often evident while interacting with them during interviews and conversations or as highlighted by their blogs and posts on social media. There is a sense of pride from project leaders and their teams about achieved project outcomes. Pride in megaprojects has been identified in prior literature. Van-Marrewijk (2007) recorded that the employees of the megaproject that he studied felt that they were constructing something unique, something not easy to realise, making them proud to be a part of something special. Such feelings of pride augment feelings of psychological ownership (Kirk et al., 2015) as actors feel proud of their organisation and its decision-making. Generally, in the case of megaprojects, the project team is very motivated to take things into their hands so that the delivery of the project would not be delayed. They talk frequently about the achievements of the megaproject and take pride in their decisions. The project team feel proud not only about the built asset but also about the decisions they took for managing external stakeholders and other constraints the asset faced.

7.2.2 Importance of the megaproject

The project team's view is that the construction of the megaproject was creating a vital piece of infrastructure. This rationality makes them regard the work they do to construct the megaproject as a service in the interests of the nation. The project team believes that the project is important for the city and often wants to be a part of it. Given the importance of the project, they defend the decisions they make regarding the planning, construction and operation of the megaproject. Thus, the decisions of the project team were motivated by their view of the importance of the project and belief that the project was the only way to solve the problems that the city faces. Their belief and motivation lead to a commitment to work on a project to achieve the desired outcomes.

7.2.3 Overemphasis on time

Project management is often described as the management of the 'iron triangle,' i.e., the management of time, cost and quality (Atkinson, 1999). In the case of megaprojects, time took the front seat, while cost and quality often faded into the background. To enable the construction, the project team compromise on costs and quality parameters. The overemphasis on time in relation to cost is seen in the way that the project team sought to make rapid progress and finish the project as quickly as possible because the asset was treated as an urgent infrastructure, as noted in the work of Ninan et al. (2022b). As a result, they opted for strategies such as extra work for stakeholders, which increased costs whilst enabling faster completion of the project. The spending of resources to achieve time objectives is a well-noted strategy in the work of Ninan et al. (2019), in which they highlight fund discretion as an important resource to enable the speedy construction of the megaproject. Along with cost parameters, quality also suffered. The asset creation team operated to a strict timeline in the midst of constant political threats and opportunism. Politically, much of the time, infrastructure assets are used strategically by political parties as objects with which to boost their appeal or, alternately, as objects of critique by oppositions. To achieve time objectives, the project team cut through the curing period (the time it takes for concrete to 'set' after being poured) in order to inaugurate the asset on the pre-committed date. Thus, the decisions taken by the megaproject were often determined by an over-emphasis on time in completion terms.

7.2.4 Acceptability of public inconvenience

The project team justified the social costs that inconvenienced the public during the construction and operation phases. They justified creating traffic diversions and disruptions in some services by saying that the people needed to adjust. The asset creation team often relied on the acceptance and adjustments of the public, rather than be active in reducing inconvenience to external stakeholders and the community. Cases of communities accepting the inconvenience caused during the construction of infrastructure asset are acknowledged in research by Yeo (1995): he emphasised that the public are able to understand long-term benefits of an asset and can even tolerate the inconvenience caused during construction phases. In the case of megaprojects, the project team take for granted that there would be public inconvenience and their decisions were based on this rationale.

These four rationalities shaped the decisions taken by the megaproject. These shared rationalities enable them to work together with common understanding and internal motivation allowing work teams to execute tasks with little monitoring or supervision (Levitt, 2011). The next section explores how these rationalities came into being and are maintained.

7.3 Source of motivations

Four sources of rationalities were observed in the megaproject that emerged from the data. These sources of rationalities were: instilling pride; shaping identity; creating urgency; and working through hardships. We discuss each of these below.

7.3.1 Instilling pride

The project team promoted strategic narratives and discourses to manage external stakeholders through social media posts and news articles. Transportation megaprojects are planned as an environmentally friendly transportation alternative and hence rely on electricity for its operation and promoted themselves as having the largest on-site solar projects. In some cases, constraints force the megaproject to opt for certain decisions; however, these are promoted and marketed in social media and news articles as the megaproject's vision to achieve something great, seeking to instil pride. Since the megaproject is colossal, the community would not know the full story behind these decisions, other than what they read from news articles or social media posts.

A community member responded in the work on Ninan et al. (2022b), "I feel proud. Thanks to *** (metro rail organization)". The project team reciprocates this sense of pride in the achievements of the megaproject and defends their decisions of the megaproject. Similarly, Dixon (2017) noted that compelling stories that interpret the iconic achievements of assets are instrumental in creating a unique selling point of the asset to the leadership team. Thus, we highlight that the strategy of instilling pride in the community by portraying the asset as 'biggest', 'largest' or 'trendsetter' can also result in the creation of a 'pride in the asset' rationality on the asset creation team. A process of instilling pride can be strategic and promotional in nature for a purpose of sharing narratives and stories about projects to known people, to attract more attention from investors and others, to get new project work.

7.3.2 Shaping identity

Megaprojects celebrates national and regional festivals and propagates them in the social media posts and news articles. For example, Ninan et al. (2022b) in their study of a metro rail project in India note social media posts of the project's celebration of regional festivals. Along with festivals, these projects celebrate celebrates national days, visits by eminent personalities to the construction sites, and Corporate Social Responsibility (CSR) activities carried out by the megaproject. Such cultural celebration aimed to make a brand iconic by associating it with significant events and festivals (Holt, 2004). Many members of the community see megaprojects as something that, through these celebrations, resonate with them as well as representing their identity. Given that the megaproject became tantamount to the identity of the city, the project team treat the asset as being of national importance and claimed their work on the metro rail asset as a kind of 'national service'. Often talented employees seek opportunities to associate with and work on the megaproject. Job attraction is mentioned as one of the effects of branding on employees by Dineen and Allen (2016). Sivertzen et al. (2013) notes that branding through social media positively relates to organisation reputation, which motivates employees to apply for a job in the organisation. The role of branding in enabling the recruitment of talented workforce has been noted previously in Tumasjan et al. (2016). With megaproject employees coming from international projects and government services, the talent-attracting effect of hegemonising strategies through branding is evident Thus, the strategy of shaping the identity of the asset for the wider community resulted in the 'importance of the asset' becoming a key rationality for the asset creation team.

7.3.3 Creating urgency

The team rationalised its decision in selecting and constructing the megaproject through strategic discourses. Megaprojects constantly propagate messages in their public outlets, such as "Cities that have created integrated, multi-modal infrastructure have effectively tackled pollution and congestion", as noted in the work of Ninan et al. (2022b). Through these discourses the megaproject identifies itself with urgent goals in public life such as reducing pollution and congestion. Thus, the asset became perceived by the community as urgent, essential and to be completed at all costs. Urgency recognises time like an arrow and since launching anything new requires commitment and intense effort, urgency is often the way to make it happen in organisations (Ganz, 2011). The sense of urgency promoted to the community through strategic messaging was similar to the project team's 'overemphasis on time' rationale. Both the community and the asset creation team believed that the asset was urgently required and so the team's decisions were made and justified according to this rationale.

7.3.4 Working through hardships

The hardships incurred during the process of construction of the megaproject are communicated through news articles and social media pages. Construction of the megaproject during the night hours is communicated through strategic discourses as 'toiling when the city sleeps'. Acknowledgment by the community of the hardships of constructing a megaproject can be related to the 'acceptability of public inconvenience' rationale of the project team. Similar to the community, the team also believed that the megaproject was being completed in the midst of many constraints for the benefit of the community and hence the community had to adjust to the inconvenience caused.

7.4 Narratives, motivations and sense of pride

The project team rationalities were picked up from strategic discourses aimed at 'cooling out' the community stakeholders. Ninan et al. (2022b) highlight that these were not messages that were initially intended to build project team cohesion but were oriented to minimising the resistance and maximising acceptance by the community in which the works were being undertaken and were part of strategic asset management. In the case of the megaproject, the project team is also part of the stakeholder environment as they were positively influenced by the narratives employed to manage the external stakeholders. All the project participants were also members of the community and they were equally influenced through the news articles and social media posts, even when they knew that there were sometimes more complex stories behind and beneath the news. Adding to this, the project team, being recruited from the community and continuing to live among the community during the duration of the project, had a dual identity as a part of the megaproject and as a part of the community. Therefore, the team was part of the social fabric of the community and was not immune to the strategic messaging of the megaproject. They are, thus, both the actor and the subject of these strategies. This broadens the definition of asset community from NGOs and community groups (Gellert & Lynch, 2003) to all stakeholders of the asset such as the asset creation team, financiers, government, etc., as all these stakeholders are part of the community and are influenced by the news articles and social media posts (Mathur et al., 2021). While the importance of customer engagement for megaproject management is recorded in Blom and Guthrie (2015), this work extends the literature by exploring the relationship between customer engagement and project team rationalities.

7.5 Conclusion

In this chapter, the motivations and sense of pride in megaprojects are discussed. We discuss different motivations in megaprojects such as pride in the megaproject, the importance of the megaproject, overemphasis on time, and the acceptability of public inconvenience. We highlight how these motivations are dependent on narratives of instilling pride, shaping identity, creating urgency and working through hardships. We also record how narratives are important for unifying both internal and external stakeholders, and necessary for the successful completion of a megaproject. Finally, the way forward to explore narratives in megaprojects, both in research and practice, is stressed upon.

CASE STUDY 7:   CHENNAI METRO RAIL, INDIA

The Chennai Metro Rail is a metro rail project in the city of Chennai in India. Chennai is one of the largest metropolitan cities in India with a population of around 10 million people. The city has been growing rapidly and the traffic volumes on the road turn chaotic during the peak hours of the day. This is because the use of public transportation within the city does not match the growth of population in the city due to the migration of people from rural to urban areas. Chennai realised that it needed a metro rail transport network to solve the issues of the existing public transportation system, to accommodate the exponential growth of population fuelled by massive urban migration to the city and to achieve the city's intended economic growth. With an aim to reduce pollution and a vision of 'moving people, sustaining growth' the Chennai Metro Rail Limited (CMRL), a quasi-government organisation, was incorporated in 2007 as a joint venture between the Government of India and the government of the state with equal equity holding. In the same year, the detailed project report (DPR) for the project was also completed. Subsequently, funding was approved in 2008 and construction activities commenced in 2009. Multiple small sections of the project were completed and opened to the public from 2015 to present; however, the whole phase 1 of the project is not yet complete during the study period.

The phase 1 of the Chennai Metro Rail project considered for this study has a planned cost of US$2.2 billion and thus satisfies the quantitative requirement of being a megaproject (cost greater than US$1 billion). The project was proposed as the answer to the traffic demands of a rapidly growing city. There was rapid urbanisation in the city, which resulted in an increase in privately owned vehicles, road congestion and consequent air quality problems. Thus, the project was conceived with multiple objectives, such as boosting the economic growth of the city and reducing pollution. The project was also aimed at providing interconnectivity with existing public networks including buses, sub-urban trains and Mass Rapid Transit System (MRTS), and an eco-friendly alternative to existing modes of transport. The phase 1 of the project consisted of two corridors of a total length of 45.1 kilometres which had both elevated and underground sections, with the majority (55 per cent) being underground. An elevated stretch of the Chennai metro rail is shown in Figure 7.1.

The project team was proactive in engaging with the external stakeholders. The metro rail organisation celebrated national and regional festivals and propagated them in the social media posts and news articles. A tweet reporting this read,

> CMRL celebrates Pongal (local festival in the region) festival at *** (name of the station) Metro Station on 7th & 8th Jan 2017.
>
> *(Tweet by official page on 6 January 2017)*

FIGURE 7.1 Chennai metro rail trains on an elevated stretch

Along with celebration of festivals and the hoisting of flags during patriotic days and their promotion in social media, the metro rail project also shared its achievements in the news media articles with the wider community, such as being one of the largest on-site solar project in India, as shown below.

> The total capacity will be six MW, which will make it one of the largest on-site solar projects in India.
>
> *(Quoted from a news article of 23 June 2017)*

The project was well received by the community and was perceived as being urgent, essential and to be completed at all costs, something evident in this community comment on Facebook:

> The sooner this stretch is declared open, it will be better for the general public.
>
> *(Quoted from the comments of a news article of 17 January 2017)*

In addition to awakening pride in external stakeholders and the resulting support, there was also a sense of pride with the internal stakeholders. The project team also frequently talked about the achievements of the megaproject and took pride in their decisions as part of infrastructure asset management.

For instance, land acquisition is a challenging process for infrastructure assets. Sometimes the land is owned by the government and allotted for construction to the metro rail organisation only for the project organisation subsequently to understand that the land is not vacant but is currently occupied by illegitimate landholders. One such instance occurred with a parcel of land near the airport that was critical for connectivity to the airport: after it was formally handed over by the authorities it was found that the airport was occupied by illegitimate landowners. Due to the proximity of this illegitimate settlement to the airport and concerns about safety, in the past the airport authority had unsuccessfully tried to evict the squatters. The asset creation team was able to evict illegal landholders and carry out the construction. The section head who was in charge of the construction remarked during an interview:

> The airport couldn't do it (evict illegitimate landholders) for 15 years, but we were able to do it in a short span.

The project team believed that the asset is important for the city and wanted to be a part of it. The positive image of the project led many people to work for the project leaving their earlier jobs. One employee said,

> I resigned a government job [government jobs are very prestigious in the Indian context] to be here … This is a big project happening in my city … I want to be part of it.

Even though the project had to acquire land, cause disruptions to the city and cut across multiple utilities, including sewers, water mains, telecommunication cables, electricity cables, there were no significant stakeholder management issues in the project. The project was awarded the "National Project Excellence Award" in 2019 for the successful completion and commencement of passenger services for the entire Phase-1 project in an extremely challenging condition while keeping public safety as a priority.

Exercise

1. How did the Chennai Metro Rail create a favourable narrative?
2. How was the sense of pride evident in the project?
3. Why did the project team share the sense of pride on social media?

Sources

PTI. (2017). Solar energy to power Chennai Metro, Clean Max bags contract. https://www.livemint.com/Politics/zfQrQjstb9to0JNRB6Yf2N/Solar-energy-to-power-Chennai-Metro-Clean-Max-bags-contract.html (Accessed on 10 August 2022).

Sekar, S. (2017). Final inspection of Metro Rail underground stretch soon. https://www.
thehindu.com/news/cities/chennai/Final-inspection-of-Metro-Rail-underground-
stretch-soon/article17048931.ece (Accessed on 10 August 2022).

Sekar, S. (2019). Rush hour in the city sees Metro trains running full. https://www.
thehindu.com/news/cities/chennai/rush-hour-in-the-city-sees-metro-trains-
running-full/article29858368.ece/amp/ (Accessed on 10 August 2022).

References

Atkinson, R. (1999). Project management: Cost, time and quality, two best guesses and a phenomenon, its time to accept other success criteria. *International Journal of Project Management*, 17(6), 337–342.

Bansal, P., Smith, W. K., & Vaara, E. (2018). New ways of seeing through qualitative research. *Academy of Management Journal*, 61(4), 1189–1195.

Blom, C. M., & Guthrie, P. M. (2015). Surveying customer perceptions of road infrastructure comfort. *Infrastructure Asset Management*, 2(4), 173–185.

Dineen, B. R., & Allen, D. G. (2016). Third party employment branding: Human capital inflows and outflows following "best places to work" certifications. *Academy of Management Journal*, 59(1), 90–112.

Dixon, M. (2017). Asset management: Making the 'unreasonable' reasonable? *Infrastructure Asset Management*, 4(1), 3–7.

Flyvbjerg, B. (2008). Curbing optimism bias and strategic misrepresentation in planning: Reference class forecasting in *practice. European Planning Studies*, 16(1), 3–21.

Frawley, J. K., & Dyson, L. E. (2018). Literacies and learning in motion: Meaning making and transformation in a community mobile storytelling project. *International Journal of Mobile and Blended Learning (IJMBL)*, 10(4), 52–72.

Ganz, M. (2011). Public narrative, collective action, and power. In S. Odugbemi & T. Lee (eds.), *Accountability through public opinion: From inertia to public action* (pp. 273–289). Washington: The World Bank.

Gellert, P. K., & Lynch, B. D. (2003). Mega-projects as displacements. *International Social Science Journal*, 55(175), 15–25.

Holt, D. B. (2004). *How brands become icons: The principles of cultural branding*. Boston, MA: Harvard Business Press.

Kahneman, D., & Lovallo, D. (1993). Timid choices and bold forecasts: A cognitive perspective on risk taking. *Management Science*, 39(1), 17–31.

Kirk, C. P., Swain, S. D., & Gaskin, J. E. (2015). I'm proud of it: Consumer technology appropriation and psychological ownership. *Journal of Marketing Theory and Practice*, 23(2), 166–184.

Levitt, R. E. (2011). Towards project management 2.0. *Engineering Project Organization Journal*, 1(3), 197–210.

Locatelli, G., Mariani, G., Sainati, T., & Greco, M. (2017). Corruption in public projects and megaprojects: There is an elephant in the room! *International Journal of Project Management*, 35(3), 252–268.

Mathur, S., Ninan, J., Vuorinen, L., Ke, Y., & Sankaran, S. (2021). An exploratory study of the use of social media to assess benefits realization in transport infrastructure projects. *Project Leadership and Society*, 2, 1–10.

Merchant, G. U. Y. (2009). Web 2.0, new literacies, and the idea of learning through participation. *English Teaching*, 8(3), 107–122.

Ninan, J., Mahalingam, A., & Clegg, S. (2019). External stakeholder management strategies and resources in megaprojects: An organizational power perspective. *Project Management Journal*, 50(6), 625–640.

Ninan, J., Mahalingam, A., & Clegg, S. (2022a). Power in news media: Framing strategies and effects in infrastructure projects. *International Journal of Project Management*, 40(1), 28–39.

Ninan, J., Mahalingam, A., & Clegg, S. (2022b). Asset creation team rationalities and strategic discourses: Evidence from India. *Infrastructure Asset Management*, 8(2), 1–10.

Ritchie, J., & Lewis, J. (2003). *Qualitative research practice: A guide for social science students and researchers*. London: SAGE.

Sivertzen, A. M., Nilsen, E. R., & Olafsen, A. H. (2013). Employer branding: Employer attractiveness and the use of social media. *Journal of Product & Brand Management*, 22(7), 473–483.

Tumasjan, A., Kunze, F., & Bruch, H. (2016). Linking employer branding and firm performance: Testing an integrative mediation model. *Academy of Management Proceedings*, 1, 14112.

Van Marrewijk, A. (2007). Managing project culture: The case of Environ Megaproject. *International Journal of Project Management*, 25(3), 290–299.

Wachs, M. (1989). When planners lie with numbers. *Journal of the American Planning Association*, 55(4), 476–479.

Yeo, K. (1995). Planning and learning in major infrastructure development: Systems perspectives. *International Journal of Project Management*, 13(5), 287–293.

Zuashkiani, A., Schoenmaker, R., Parlikad, A. K., & Jafari, M. (2014). A critical examination of asset management curriculum in Europe, *North America and Australia, IET/IAM Asset Management Conference*, London.

8

NARRATIVES OF THE FUTURE

8.1 Introduction

Narratives, and more specifically project narratives, shape and change the future. Project narratives create future actions. In previous chapters we discussed how project narratives become mobilised, crafted, managed, maintained and promoted to achieve the desired state in the future. In this chapter we focus more on what makes a project narrative, what future brings us and the role of project narratives in it. We project our vision on narratives of future in this chapter.

8.2 What makes a project narrative?

In the first chapter we outlined key distinguishing characteristics of narratives from other forms of discourses, i.e. stories. Throughout the book we have discussed the nature and role of project narratives, their different types and how these are mobilised, crafted, managed, used maintained and promoted. As discussed previously, there are ante-narratives, what come before a coherent and persuasive project narrative, which are not yet fully formed. Ante-narratives are bets that can disrupt and transform project narrative thought. Not all project narratives are successful. We suggest that there are key features that makes a project narrative to succeed which we discuss below.

8.2.1 Future-oriented nature of project narratives

Project narratives are about future; that is, they are future-oriented. A dominant project narrative is about what a project aims to achieve mobilising human and financial resources in order to achieve the desired outcomes. Ante-narratives that

DOI: 10.1201/9781003248378-8

are about past and present and do not project future are likely be unsuccessful and not to become a project narrative. From a pool of ante-narratives one future-oriented may become dominant that is crafted and maintained throughout a project lifecycle. Project narratives are about future vision. They are about a process of projecting the desired future.

8.2.2 Project narratives are connected with grand narratives

Project narratives are connected with grand narratives. Project narratives address bigger industrial, national and global agendas. In Chapter 2, we showed examples of megaproject narratives and their connectivity with grand narratives. According to Fenton and Langley (2011), broader institutionalised "grand narratives" can be distilled from the analysis of sets of texts at particular times in history, and that provide meaning for practitioners in their organisations. Research has demonstrated the connectivity between narratives at national and international levels (Sergeeva & Lindkvist, 2019), at government and project-based firm levels (Sergeeva & Winch, 2020). The key grand narratives of the future are: global net human-caused emissions of carbon dioxide need to fall by about 45 percent by 2030 reaching net zero by 2050. The connectivity and alignment between project narratives and grand narratives provides coherence and conformity important for achieving common goals. A project narrative that connects with the grand narratives is likely to succeed and become sustained over time. For example, the 'green' agenda of the HS2 was repeated multiple times:

> As someone who cares deeply about the environment, the opportunity to dramatically expand rail, a greener form of transport than aviation or road is very exciting indeed. This investment will help people to choose trains over cars, reduce carbon emissions and provide a rail system that is faster, more reliable and greener.
>
> *(Quoted from the news article 'Train service from Aylesbury via Winslow to Milton Keynes gets the green light', dated 16 July 2012)*

Project narratives get repeated in order to (re-)emphasise the message.

8.2.3 Strong project narratives to overcome counter-narratives

Project narratives need to be strong to overcome resistance to change. There are likely to be conflicting views and counter-narratives by protesters and those who are not in favour of a projects (Ninan & Sergeeva, 2021). Hence project narrative needs to be strong in overcoming such resistance. The stronger project narrative is likely to have less attention paid to counter-narratives. Counter-narratives would become less important when there is a strong dominant project narrative.

8.2.4 *Project narratives inspire people*

Project leaders craft and communicate a project narrative that inspires employees, excites partners, attracts customers and engages influencers and, perhaps most importantly, investors. The project narrative is used to explain why the project exists and what makes it unique, the value and relationships it creates and communicates these to both internal project team members and external stakeholders (Ninan et al., 2021). This is why statement from leaders are vital for generating belief in the project – as Sir Tim Smit (2002) said of the Eden Project – projects are like Tinkerbell; they only exist if you believe in them. The case is presented at the end of this chapter.

The inspiring nature of project narratives attract attention from the public, and help to win new projects. Inspiring project narratives engage people, persuade people and promote achievements and stories of success. A sense of pride and achievement makes a project narrative inspiring for others and future work. As we discussed in the previous chapter, motivation and sense of pride are very important in project work.

For example, during the early stages of the HS2 project, the secretary of state for transport created a narrative of the need for project by highlighting that it will transform transport in the country and provide numerous benefits:

> I am excited about the possibilities that HSR [High Speed Rail 2] has to transform transport in this country for the better – providing environmental benefits, encouraging investment and boosting business and jobs.
>
> *(Quoted from the news article 'High speed rail must have Easytrain prices says Transport Secretary', dated 30 December 2009)*

Project narratives show excitement about a project and inspire and encourage people to commit to project work.

8.2.5 *Project narratives create project identity and image*

Project narratives and narrating play pivotal roles in creating project identity and project image. It is through project narratives that project identity or project DNA becomes created and communicated among the team and more widely. Future-oriented project narratives create shared vision among the project team to work together to achieve the desired future. Project narratives connected to grand challenges create a sense of working together in a project towards achieving goals. Strong and inspiring project narratives are important part of project identity. Project narratives are essential for project image or project branding that becomes crafted and promoted widely.

For example, Ninan et al. (2019) studied the branding practices of a megaproject using Foucault's governmentality theory. The megaproject used narratives through social media to create a brand image of the project as beneficial for the community resulting in support for the project. For example, the branding effect of narratives can lead to community support for project activities. Project image narrative helps to overcome resistance to change, and turn protesters into supporters.

8.3 Projecting narratives of future

As we discussed in previous chapters, the future-oriented narratives are visioning or projecting narratives. The future is inherently unknowable, yet projecting involves an attempt to shape that unknowable future in alignment with the past and present. The projecting nature of narratives means that they potentially shape and change the future (Ninan & Sergeeva, 2022; Sergeeva & Winch, 2021). A process of narrating is a projecting process of the future. It is an essential process of strategising and decision-making under conditions of uncertainty (Kay & King, 2020).

Narratives tend to be deliberately used to create desired future. Narratives generate future actions. Narratives about the desired future play a crucial role in branding a project (Ninan et al., 2019). It creates an image for a project to be promoted to external stakeholders and more widely. Project narratives address future trends; they are relevant. Narratives such as low-carbon projects, the sustainability of projects, innovation in projects and digitisation in projects are relevant.

Narratives of the future create goals and opportunities for findings ways of achieving them. Narratives about net zero economy by 2030, 2050, 2070, responding to the challenges of climate change and living sustainably are key narratives of the future adopted by the ecosystem of industries, firms, projects and individuals who need to become committed to deliver and achieve goals and desired future. The act of commitment is an investment in projects that address these priority narratives. Industrial decarbonisation projects will unlock a low-carbon future and respond to the challenges of climate change. Intelligent technologies enable to meet commitments to society on its journey to a net zero economy. Actors play pivotal roles in narrating their responses to narratives of a net zero economy and challenges of climate change demonstrating real commitments and actions.

Project leaders are projecting narratives of future. They think about the way forward and what is yet to come. The future provides new ideas, new opportunities and inspirations for project work. The world becomes transformed through projects. Narratives in projects play important roles in transformation. Narratives become commitments and actions towards achieving goals.

8.4 Future directions

In this book, we highlight the importance of narratives in megaprojects and describe the narrative instruments and narrative processes used to craft and maintain narratives. There is a need for more research on narratives and changes to practices.

Narratives are particularly important in project settings as they are temporary organisations that bring multiple diverse stakeholders together to achieve the common goal (Sergeeva & Ninan, 2022). We outline key features of what makes a project narrative. Narratives of the vision of a project create an identity for the internal stakeholders which helps in collaboration and can create a positive image for external stakeholders which helps in garnering support for the project, both of which increase the likelihood of project success. In this book we discuss some of the ways to craft narratives such as stories, labels and comparisons, and some ways to maintain project narratives such as repeating, endorsing, humourising and actioning upon. We contribute to a theory on project narratives. Our work has important implications for practice.

There is still more research work that needs to be done to understand the practice of managing, leading and organising projects through narratives. While some organisational theories such as sensemaking, sensegiving, social identity theory, and organisational power theories are considered in project settings, other theories, such as attribution theory (Vaara, 2002), institutional theory (Skoldberg, 1994), practice theory (Carlsen, 2006) etc., can be employed to make sense of narratives in projects.

Among the research methodologies that can aid researchers in exploring project narratives are narrative interviews, participant observations and online naturalistic inquiry, such as the study of social media, news articles and project newsletters. While narratives in project studies are usually associated with written texts or spoken words, other forms of communication, such as visual, audio and even actions, can be considered. A study on video narratives can significantly inform project management practice since they have a richer ability to account for emotions in the form of changes in tones, calmness and anxiety (Huy, 2002). Many projects, such as the Tideway project in the UK, upload short videos in YouTube to disperse their message to a wider audience. These short videos capture people's experiences of their everyday work. Future research can study the role of these videos in project organising and narratives. Of further interest is how posted videos of current projects by project leaders enable them to win next projects. Similarly, a study of 4D CAD models capture how project evolves and coordination with other stakeholders (Datta et al., 2020). Other avenues, such as micro-stories by the project members of the work they do (Fenton & Langley, 2011), can significantly improve our understanding of the project management career.

The forms of narratives are evolving from spoken and written to video forms. The technical capability of mobile devices privilege digital media (e.g., video) and multimodal content (e.g., image, video, sound) over traditional written text (Frawley & Dyson, 2018). With the advent of social media, all stakeholders have the tools with which to produce and share their own multimedia culture and meanings. The user-generated content platforms support a widespread participatory culture where non-experts have the ability to create and share new content online (Merchant, 2009). They have a voice now which can be easily transmitted and heard, and the implications of this voice of the society on the megaproject narratives and outcomes have to be studied. In the modern era, many narratives in megaprojects are in the digital world. Researchers have to look for narratives in these digital sources to understand society and how they organise, to support or disrupt megaprojects. Research on narratives need to observe narratives and practices in the digital space to understand the insider's world of meaning (Ritchie et al. 2003). Researchers can explore these 'new ways of seeing' (Bansal et al., 2018) in megaproject settings. While researching digital data for narrative research in megaprojects, researchers have to be cautious of potential pitfalls of employing online naturalistic data, such as the poor representativeness of data, the lack of guidelines, and the traceability of data (Ninan, 2020). To understand project narratives in megaprojects in the twenty-first century, we also have to study the practice of projects in the online environment (Ninan, 2020). Many conversations relating to project happens on Facebook, Twitter, LinkedIn, WhatsApp, Instagram and news media articles. Social media provides an opportunity to the project community and empowers the marginalised by providing them with an audience for their stories (Vaara et al., 2016). In social media, different forms of data, such as text, picture, video, and so on, are often interwoven (Ninan et al., 2019). The archive of digital media enables researchers to study narratives longitudinally. Even retrospective data relating to a project can be retrieved and analysed for the role of narratives. Additionally, project narratives in real time can be studied to understand the convergence and divergence of meaning, such as in the case of project benefit realisation (Mathur et al., 2021). Therefore, to better understand the process of narrating in project organising, more longitudinal, multi-methods, multi-theoretical research is called upon, which will in turn help us to understand and improve project management practice.

The practice of stakeholder management in megaprojects also have to improve significantly as more and more projects experience disruptions due to their inability to engage external stakeholders. It is generally perceived that public projects should not attempt narratives and marketing. On the contrary, as Ganz (2011, p. 284) notes:

> If we do public work we have a responsibility to give a public account of ourselves: where we came from, why we do what we do, and where we think we're going. In a role of public leadership, we really don't have a choice about telling our story of self. If we don't author our story, others will. And they may tell our story in ways that we may not like, not because they are malevolent, but because others try to make sense of who by drawing on their experience of people whom they consider to be like us.

Most megaprojects have a public relations team and have a presence in social media and digital spaces. Marketing videos and advertisements can be seen across megaprojects. In the case of the Westconnex project in Sydney, Australia, the project even sponsored the local football team to resonate more with the identity of the city. The project team should also reach out to leaders, celebrities, and other people with referent power to endorse the different narrative instruments. Humourising and actioning of the narrative instruments can also help stabilise the project narratives. Projects adopt these strategies to create a stable narrative. However, the use of narratives should not be a substitute for engagement as stakeholder consultation and mobilising narratives should go hand in hand with listening to stakeholder concerns to maximise the megaproject's value for the society.

8.5 Conclusion

In this chapter we discuss what makes a project narrative, outlining key distinguishing features. We projected our vision on narratives of future. Thinking about future narratives inspires ideas and creates new opportunities. Some examples of future narratives and how projects respond to them are provided. We set the research agenda into project narrative work including directions of research, methodologies and research methods.

CASE STUDY 8: EDEN PROJECT, UNITED KINGDOM AND WORLDWIDE

The Eden Project (www.edenproject.com) in Cornwall is one of the most successful UK Millennium projects. It opened in March 2001 to provide an outstanding experience around one million visitors per year – double the number originally envisaged in 1997. Constructed in a redundant south-facing china clay pit, a large covered biome provides a humid tropical environment, while a smaller one provides a warm temperate environment, totalling between them 2.1 hectares. The cool temperate environment is in the third, uncovered, outdoor biome. An education centre – The Core – opened in 2005. The driving forces of the early days of the project were Tim Smit, who had rescued and opened to the public the nearby Lost Gardens of Heligan in 1992, and Jonathan Ball, a successful local architect. The project presented an enormous range of challenges and provides a vivid example of the power of what Smit calls "telling future truths".

In 1993, the UK government established the National Lottery to fund, amongst other good causes, the Millennium Commission. Its purpose was to celebrate the forthcoming second millennium with calls for "a scientific or engineering project that becomes one of the wonders of the third millennium". Amidst disparaging remarks about Æthelred, the English king at the time of the first millennium, the call was for a future-orientated initiative that would excite and inspire. A number of projects of various kinds were supported by the Millennium Commission with the aim of celebrating the coming of the third millennium and leaving a lasting legacy. The most successful of the 12 larger Landmark projects was The Eden project.

The idea for Eden was distilled from a conversation over a bottle of whisky in a West Country farmhouse kitchen one night in May 1994. Smit and Ball complemented each other with Smit's horticultural expertise and fluid ability to articulate compelling narratives, and Ball's architectural expertise and extensive networks amongst the higher echelons of both Cornish and London society. These latter connections included official roles for the Royal Institute of British Architects, and membership of one of the grander gentleman's clubs.

The project mission they generated in the autumn of 1994 from the initial whisky-fuelled idea was:

> To create under one roof a range of natural habitats found on planet Earth ... An international resource designed for research, education, and public enjoyment to herald the new Millennium, bequeathing a gift of incalculable value to those who will follow us ... our hope for and belief in the future.

Funded by pump-priming money from the local government sources, a mix of Smit, Ball, other local players, and horticulturalists energetically shaped this project mission. The project was possibility and not fantasy to due to the launch of the Millennium Commission in February of that year. They captured early ideas in a variety of architectural sketches in both plan and elevation, sometimes prepared on restaurant menu cards. This resulted in Eden being submitted as the UK entry to the architectural Venice Biennale in 1996.

An outline proposal was submitted in April 1995. The first paragraph of that submission read:

> The concept of the Millennium is rooted in recognition of that significant midnight when we look backward to the past and forward to the future simultaneously. Its social value lies in concentrating our minds on past achievements, present problems and future possibilities. Any project designed to mark this transition should excite interest, understanding and involvement in shaping a desirable future.

This outline proposal was turned down as underdeveloped, but this did not faze Smit. Upon receiving the news, he said to Ball:

> we're going to bluff it out. We're going to tell everyone that we have caught their [the Millennium Commission's] imagination and have been asked to work it up some more. And what's more, we're not going to take no for an answer.

A significant re-think was required and the team decided to assemble some of the leading players in the UK construction industry to add credibility to their efforts. However, funds were very tight, and so these players were recruited through Ball's personal contacts on the basis that they would not be paid unless the project were successfully funded. Remarkably, they agreed to participate.

The architects, Nicholas Grimshaw and Partners, worked on developing the design concept. They soon realised that their original idea (a reprise of Grimshaw's Waterloo International Terminal) would not work propped against the side of the clay pit, because the structure was too heavy for the span and the ground too uneven and continually changing due to the continued working of the pit for clay. The inspiration for Grimshaw's final design was a soap bubble that can mould itself to whatever surface it alights upon; their technical solution a geodesic dome. The erection of the structure on the 858 m-long ground beam required the largest free-standing scaffold in the world, followed by the installation of the cladding panels by abseilers. Civil engineering works included moving 800,000 m³ of fill by the construction manager McAlpine JV. This consisted of Sir Alfred McAlpine plc and Sir Robert McAlpine Ltd., who came together for the first time since the firm had split in 1940 due to a family dispute because it was "the ultimate construction project".

Ball managed to convince all of these firms, together with some of the leading international consultancies, such as Ove Arup and Davis Langdon, to work for free in order to develop the design while Smit and the team then lobbied the Millennium Commission. The Commission did not fund development work prior to full bids, and so it was not obvious anything was amiss and the team struggled on private donations and small grants. By mid-1996, the lobbying achieved results and Eden was back in the competition with a submission due in December with a budget of £74.3 million. The news that Eden had been successful was announced in May 1997, and so the McAlpine JV was notified as preferred bidder for the delivery phase of the project in the following month. The relationship was reinforced by appointing a director from Sir Robert McAlpine Ltd. to the Eden Board in 1998. This relationship would be of enormous benefit later in the construction phase when the project nearly ran out of cash owing the JV millions and the McAlpine director steadied the boat by saying "we're still here".

Funding came from a wide variety of sources – Millennium Commission funds only provide 50 per cent of the total capital requirement of nearly £80 million. Smit's credibility with the success of Heligan secured seed corn funds from the county (Cornwall), local charities and private interests. The ability of Ball to network both locally within Cornwall and nationally garnering enthusiastic commitment was impressive, mobilising the right people to solve difficult problems – particularly those associated with finding the other half of the funding for the project. These skills encouraged the head of a neighbouring county, Somerset, to give their public backing to the Cornwall project for European Commission Structural Funds at his own county's loss. At the formal signing of the legal agreements for the finance Eden's legal lead noted that

> the most extraordinary thing about it all was that we'd persuaded such a wide group of people, many of whom would have found it easier to walk away, to stay at the table and find a reason for saying yes.

With the funding announcement, the project reached a turning point and Smit asserted that:

> There comes a time in all great ventures when the talking has to stop. We'd created the constituencies, we'd talked the hind legs off donkeys, we'd been snake-oil salesmen with attitude and a dream to peddle, but turning a dream into a reality needs iron in the soul, money in the bank, and military organisation.

Finally, the clay pit was purchased in October 1998, and the construction contract signed in January 1999. By this time, the McAlpine JV had worked for nearly two years without a contract, as had most of the consultants. Intensive construction on site started in February 1999, and the complete facility opened in March 2001, ahead of schedule and to budget. Eden is a remarkably successful project; Smit ascribes this success, fundamentally, to "the act of faith that enabled so many people to sign up to Tinker Bell Theory was a testament to the Spirit of Eden taking hold". In that sense, "Eden was never about plants and architecture, it was always about harnessing people to a dream and exploring what they are capable of".

Eden Project International is now developing new projects across the UK and worldwide: Eden Project North in Morecambe, Lancashire; Eden Project Qingdao, China; Terra, the Sustainability Pavilion, Expo 2020; Eden Project Colombia, South America; Eden Project Anglesea, Australia; Eden Project Dundee, UK; Eden Project Foyle, UK; Eden Project, New Zealand; Eden Project in Portland, UK; and Eden Project in Graniteville, USA:

> New Edens are our global response to the planetary emergency. The international and national destination projects are being developed with teams from the different areas, responding to local themes and needs.

The Eden Project shares stories through horticulture exhibits, arts and culture programmes, community initiatives and education work. They tell stories in forms of videos, photos and social media. The Eden narrative of a future is to inspire people and transform the world:

> We need to inspire citizenship over consumption in order to care for this planet, Spaceship Earth, our only home, that provides us – and all life – with fresh air, clean water, fertile soil, rich biodiversity, a stable climate and an awesome recycling system/Exploring how it all interconnects can transform our understanding of the world and help us to see, how together, our actions can make a difference. Eden us a beacon of hope.

Exercise

1) What role did faith in the future play in the shaping of the Eden Project?
2) How was the change from project shaping to project delivery characterised by Smit?
3) How the future is projected in the Eden case?
4) What are the key learning points from this case?

Sources

Ball, J. (2013). *The other side of Eden*. Kernow: FootSteps Press.
Smit, T. (2002). *Eden*. London: Transworld Publishers.
Smit, T. (2020). *Eden*. London: Transworld Publishers.
The original case is published in Winch, G. M., Maytorena-Sanchez, E., & Sergeeva, N. (2022). *Strategic project organizing*. Oxford University Press.

References

Bansal, P., Smith, W. K., & Vaara, E. (2018). New ways of seeing through qualitative research. *Academy of Management Journal*, 61(4), 1189–1195.

Carlsen, A. (2006). Organizational becoming as dialogic imagination of practice: The case of the indomitable Gauls. *Organization Science*, 17(1), 132–149.

Datta, A., Ninan, J., & Sankaran, S., (2020). 4D visualization to bridge the knowing-doing gap in megaprojects: An Australian case study. *Construction Economics and Building*, 20(4), 25–41.

Fenton, C., & Langley, A. (2011). Strategy as practice and the narrative turn. *Organization Studies*, 32(9), 1171–1196.

Frawley, J. K., & Dyson, L. E. (2018). Literacies and learning in motion: Meaning making and transformation in a community mobile storytelling project. *International Journal of Mobile and Blended Learning*, 10(4), 52–72.

Ganz, M. (2011). Public narrative, collective action, and power. In S. Odugbemi, & T. Lee (eds.), *Accountability through public opinion: From inertia to public action* (pp. 273–289). Washington, DC: World Bank Publications.

Huy, Q. (2002). Emotional balancing of organizational continuity and radical change: The contribution of middle managers. *Administrative Science Quarterly*, 47(1), 31–69.

Kay, J., & King, M. (2020). *Radical uncertainty: Decision making for an unknowable future*. London: The Bridge Street Press.

Mathur, S., Ninan, J., Vuorinen, L., Ke, Y., & Sankaran, S. (2021). An exploratory study of the use of social media to assess benefits realization in transport infrastructure projects. *Project Leadership and Society*, 2, 100010.

Merchant, G. (2009). Web 2.0, new literacies, and the idea of learning through participation. *English Teaching: Practice and Critique*, 8(3), 107–122.

Ninan, J. (2020). Online naturalistic inquiry in project management research: Directions for research. *Project Leadership and Society*, 1(1), 1–9.

Ninan, J., & Sergeeva, N. (2021). Labyrinth of labels: Narrative constructions of promoters and protesters in megaprojects. *International Journal of Project Management*, 39(5), 496–506.

Ninan, J., & Sergeeva, N. (2022). Mobilizing megaproject narratives for external stakeholders: A study of megaproject narrative instruments and processes. *Project Management Journal*, 53(5), 1–21.

Ninan, J., Clegg, S., & Mahalingam, A. (2019). Branding and governmentality for infrastructure megaprojects: The role of social media. *International Journal of Project Management*, 37(1), 59–72.

Ninan, J., Mahalingam, A., & Clegg, S. (2021). Asset creation team rationalities and strategic discourses: Evidence from India. *Infrastructure Asset Management*, 8(2), 1–10.

Ritchie, J., Lewis, J., & Elam, G. (2003). Designing and selecting samples. *Qualitative Research Methods*, 77–108.

Sergeeva, N., & Lindkvist, C. (2019). Narratives of innovation that address climate change agenda in the construction sector. In M. I. Havenvid, A. Linné, L. E. Bygballe, & C. Harty (eds.), *The connectivity of innovation in the construction industry* (pp. 288–292). Oxon, UK: Routledge.

Sergeeva, N., & Ninan, J. (2022). Project narratives: Directions for research. In G. M. Winch, M. Brunet, & D. Cao (eds.), *Research handbook on complex project organizing* (in press). https://www.e-elgar.com/shop/gbp/research-handbook-on-complex-project-organizing-9781800880276.html

Sergeeva, N., & Winch, G. M. (2020). Narrative interactions: How project-based firms respond to Government narratives of innovation. *International Journal of Project Management*, 38(6), 379–387.

Sergeeva, N., & Winch, G. M. (2021). Project narratives that potentially perform and change the future. *Project Management Journal*, 52(3), 264–277.

Skoldberg, K. (1994). Tales of change: Public administration reform and narrative mode. *Organization Science*, 5(2), 219–238.

Vaara, E. (2002). On the discursive construction of success/failure in narratives of postmerger integration. *Organization Studies*, 23(2), 211–248.

Vaara, E., Sonenshein, S., & Boje, D. (2016). Narratives as sources of stability and change in organizations: Approaches and directions for future research. *Academy of Management Annals*, 10(1), 495–560.

INDEX

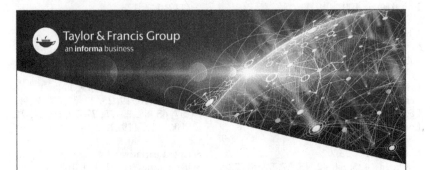

Printed in the United States
by Baker & Taylor Publisher Services